PEAT STRATIGRAPHY AND CLIMATIC CHANGE

PEAT STRATIGRAPHY AND CLIMATIC CHANGE

A palaeoecological test of the theory of cyclic peat bog regeneration

K.E.BARBER
University of Southampton

A.A.BALKEMA / ROTTERDAM / 1981

To Jane
and to my parents

ISBN 90 6191 087 0

Printed in the Netherlands

CONTENTS

PREFACE

Peat bogs are unique plant communities because of the way in which their vegetation is preserved over thousands of years through the formation of the peat itself, the partially decayed remains of the plants. In very favourable conditions such as exist in wet, actively accummulating raised bogs, the main constituents of the vegetation, bog mosses of the genus *Sphagnum,* are preserved virtually intact. This phenomenon is not only of great interest to the botanically minded however. The term 'raised bog' describes the way in which the peat mass has developed from the lake muds, fen peats or whatever preceded it, all dependent on ground water inflows, to become isolated from this mineral water influence and thus dependent upon rainfall to maintain the wet conditions necessary for active bog growth. This situation is naturally a complex one, an interaction of species with different water requirements, rainfall and its annual distribution, evapotranspiration, runoff, etc., but it has been appreciated for some time that raised bog peat can be viewed as 'proxy data' recording climatic variations.

Added fascination is given to the study of peat bogs by the extreme simplification of their plant communities imposed by the acidity of the peat bog environment. Not only do *Sphagnum* mosses hold a great deal of water because of their peculiar structure – the leaves comprise pigmented cells separating empty 'hyaline' cells which hold the water – but Sphagna possess a great 'cation exchange capacity', which means they are capable of extracting the minute quantities of nutrients in the rain which fall upon them, giving up hydrogen ions in exchange. By this means they actively acidify the local environment and how potent a process this is has been shown by Walker (1970). It used to be thought that the climax community of the hydroseral succession in most of north-western Europe and similar climatic zones would be a mixed deciduous forest, dominated by oak, as a result of the infill of lakes by mud and peat. Walker showed that once aquatic Sphagna had entered the hydrosere the trend was inexorably towards a raised bog climax community.

Another point of interest is that in no other plant community does the vegetation itself form the ground surface. In grasslands, woodlands, sand-dunes and salt-marshes the ecologist is dealing with the interface between organic and inorganic elements and the variations of slope, aspect, drainage and mineral content of the soil demand a lot of attention. On a raised bog, particularly in the central plateau areas, the micro-relief and hydrology of the surface is dependent upon the relative rates of growth and decay of the plants themselves, and, of course, on climate.

To this catalogue of interesting attributes one must add that of the preservation of microfossils, both those which inhabit the peat such as rhizopods, and those which rain out upon it, particularly pollen grains, and the way in which ombrotrophic peat acts as a collector of various mineral particles and chemicals. The pollen content of peat is one of the twin pillars of palaeoecology as practised in Europe, the other being pollen from lake muds, more favoured in the Americas. It is no part of this study to argue the relative merits of pollen analysis from bogs as against lakes but as Oldfield (1970) and others have recognised there are fewer uncertainties regarding the provenance of pollen in peat than there are in lake muds. The gratifying correlation of the pollen diagrams used in this study, and the use of peat profiles from Bolton Fell Moss as a record of particulate and radionuclide fallout, highlight the potential of peat bogs in these fields. Palaeoenvironmental work on both kinds of deposit in combination can lead to new insights and it is to be hoped that the comparative neglect of bogs in North American work will soon be redressed.

It will be apparent from the acknowledgements that this work is a slightly modified version of the author's doctoral thesis and I am grateful to the publishing house of A.A.Balkema, and to Dr W.A. Casparie, for the opportunity to publish my results in this way. I also wish to thank the University of Southampton for a loan for the making of photographic blocks in this publication.

K.E. Barber

Southampton
Christmas 1979

ACKNOWLEDGEMENTS

In any piece of work which extends over several years one is helped by numerous friends and colleagues in a variety of ways and it is no easy matter to recall all their aid and support – so to those omitted from this list, my apologies.

This project was begun during the tenure of a Natural Environmental Research Council studentship held at the Department of Environmental Sciences, University of Lancaster. I am grateful to these bodies for the chance to pursue an interest sparked off by Dr Keith Crabtree while I was an undergraduate of the Department of Geography at Bristol University, and to Professors F. Oldfield and G. Manley, as my primary supervisors at Lancaster, for much advice and encouragement, and for providing me with a magnificent research site. I also benefitted from the advice of Professor C.D. Pigott at Lancaster, and from the facilities extended to me by the Lancaster department and its later head, Professor Hunter. A grant from the Turner & Newall Research Fund of Lancaster University allowed me to join the International Phytogeographical Excursion to Scotland in 1968 and thereby gain valuable experience with Professor R. Tüxen, Dr David Bellamy and other phytosociologists. A further grant enabled me to visit Ireland to study peat stratigraphy in the company of Professor Frank Oldfield, then at the New University of Ulster, and Bob Hammond of An Foras Taluntais. The first radiocarbon date from Bolton Fell Moss was also paid for by the University of Lancaster.

I am most grateful to Dr Meybus Geyh and colleagues at the C^{14} Laboratory of the Niedersächsisches Landesamt für Bodenforschung, Hannover, for providing the further nine radiocarbon dates free of charge. I must also thank the Nature Conservancy Council (Merlewood Research Station) and the County Record Office, Carlisle, for allowing me free access to their files, and Dr O.W. Heal of Merlewood for introducing me to the study of rhizopods. The late John Mackereth, Jean Lishman and other staff of the Freshwater Biological Association at Windermere kindly helped with the iodine analyses. I have benefitted from discussion and correspondence with Professor Hubert Lamb, Professor Sir Harry Godwin, Professor Donald Walker, Dr W.A. Casparie, Dr K. Tolonen and many others.

This research would not have been possible without the fullest co-operation from Major Peter Bryceson, Lt-Col. J. Hooper, and all their staff of the Boothby Peat Company at Bolton Fell Moss; a grant of £20 from the company towards the cost of this work is also gratefully acknowledged.

Most of the analytical work reported in this work was done while I was a member of staff of the Geography Department, University of Southampton, headed by Professors J.H. Bird and K.J. Gregory. It is a pleasure to thank all my colleagues for their encouragement, and to thank Mr A.S. Burn and his cartographic staff for their high-quality draughtsmanship and reproduction of many of the figures in this work.

Dr S.B. Chapman of the Furzebrook Research Station, Institute of Terrestrial Ecology, and Visiting Fellow of the University Department of Biology, supervised the later stages of this research. His constructive criticism and advice are greatly appreciated.

I am grateful to Blackwell Scientific Publications, the Cambridge University Press, Professor H.H. Lamb, Dr W.A. Casparie and the Controller of Her Majesty's Stationery Office for permission to reproduce copyright material.

The production of the original typescript was organised by Mrs R. Flint and Miss J. Coull of the Geography Department; I am most grateful for the accuracy and neatness of Jennie Coull's typescript.

To my family I am more than grateful, especially to my wife Jane, for unfailing support and encouragement.

1. INTRODUCTION

This study was conceived as an attempt to examine the relationship between bog growth and climatic change more closely than hitherto and to test a theory of bog growth which, though privately doubted by some ecologists, still holds an important place in the literature. This theory, known variously as the 'regeneration cycle theory' or the 'hummock-hollow theory' was formulated in the early years of this century by Swedish workers and thereafter became widely established by the 1930's. The theory attempts to account for observations of both the surface microtopography and species mosaic of modern bog surfaces, and of certain stratigraphic structures in peat which must reflect in some way the surface situation. An idealized view of this 'regeneration cycle' is given in figure 1. The vegetation succession shown is a hydrological gradient from the bottom of a wet hollow, often with open water, through to the top of a drier hummock. That the various species mentioned are zoned in this way is confirmed by many observations (Osvald 1923, Tansley 1939, Hansen 1966) but this zonation must not be interpreted too strictly as many 'hummock' species may be found growing in or on the edge of pools (Ratcliffe & Walker 1958), and some species have two different 'ecads', or ecological variants, according to water level (Green 1968).

The wavy lines in figure 1 represent the bog surface at a number of different stages of peat build-up. At stage 1 the bog microtopography consists of a hummock flanked by two hollows; according to the cyclic regeneration theory, the hollows are then supposed to accumulate peat rapidly, as shown by the rising surfaces (pecked lines) while the hummock top, capped with a supposedly moribund community of *Calluna/Eriophorum* and lichens, remains at the same absolute level, either degenerating or at least not actively accumulating peat. Upward growth of the *Sphagnum*-rich hollow communities eventually raises these areas to form two new hummocks, at the same time flooding the low-lying area occupied by the old hummock – stage 2 on figure 1: the cycle is repeated – stages 3 and 4 – so that the bog surface is at any one time a mosaic of hummocks and hollows in various stages of infill, degeneration or building-up. The stratigraphic expression of this cyclical process of peat accumulation is then supposed to be seen in a cut section of peat as a series of len-

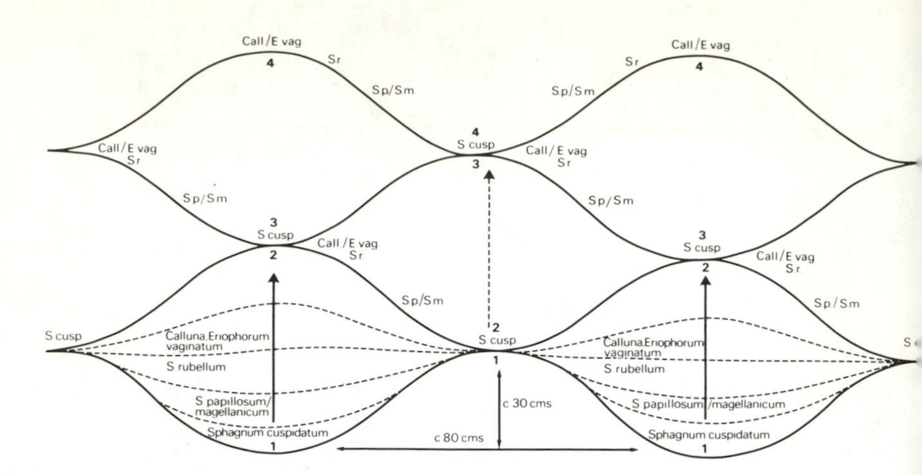

Figure 1. Idealized cycle of the regeneration complex.

ticular structures with lenses of lighter, less humified peat, corresponding to the hollow phase, and lenses of darker, more humified peat originating from the hummock phases (von Post & Semander 1910).

At the time they were formulated these ideas were in keeping with contemporaneous work in the natural sciences which was very much concerned with evolution, cyclicity and dynamism, and so this theory seems plausible and was widely accepted. It was not deliberately tested, in the Popperian sense, until some 40 years or more had passed and even so the results were not conclusive (Walker & Walker 1961). Since then research on peatlands which may have shed some light on the mechanism of bog growth has been hampered either because it was concerned mainly with the present vegetation surface and its immediate precursors – and therefore did not cover a long enough time-span (peat accumulation averages 1 cm every 20 years in raised bogs, according to Walker 1961) – or because, if it was a stratigraphic study, there was no possibility of correlating the changes in peat-type with independently-known climatic changes, either because the peat chronology was unknown, or because the surface peat had been destroyed by man. (Since the work for this study was completed a distinct link between climate and peat formation has been demonstrated by Aaby (1976), but this only touches on the hummock-hollow theory incidentally.) Indeed in most European countries the raised bogs have been so exploited and destroyed for fuel and agriculture over centuries that even if good peat sections exist it is very difficult to determine exactly where they are in relation to the former bog plane, the area of the 'Regeneration Complex'.

We have a situation, therefore, where in Walker's words (1961): 'despite the fact that their vegetation must represent some of the most natural left in the

British Isles, our knowledge of the dynamic ecology of bogs is very incomplete'. Although there have been a number of advances in research since 1960 the nature of the relationships, between climate, vegetation and peat accumulation are still unclear and their clarification is one of the main aims of this study, together with an examination of the validity of the cyclic hypothesis.

The approach taken in this study is avowedly historical for it is doubtful whether a surface vegetation study could ever give a clear answer to a dynamic ecological problem extending over a timescale of several centuries. But palaeo-ecological research has its own problems: peat formation is a diagenetic process, one which involves the translation of a living vegetation assemblage into a 'death assemblage' and then a 'fossil assemblage'. Reconstructing the main features of the 'life assemblage' is not too difficult but there are constituents of the bog vegetation which are not represented in the macrofossil analyses. Similarly the problems inherent in the use of pollen analysis and of climatic records and radiocarbon dating must be considered and for these reasons the methods used are examined in some detail.

The investigation took place mainly at one site, Bolton Fell Moss, Cumbria, with supporting observations from elsewhere. Logically no more than one site is necessary for the falsification of the regeneration theory, as long as the bog is ombrotrophic and comparable in species composition with other bogs in north-west Europe. Practically, however, it also proved impossible to work on more than one site both because of the time-consuming nature of the palaeo-ecological techniques used and because no other suitable site existed in the area. Indeed it is fortunate that most of the field investigation took place during 1968-70 when hand-cutting of the peat, which is used for garden moss-litter, was the rule; since then machine cutting has taken over entirely and the resultant compression, distortion and smearing of the peat stratigraphy would now make such investigations much more difficult. As it was many hundreds of metres of peat sections could be examined without any preparation and this was felt to be important in ensuring that representative sections were used for detailed study.

The review of previous research is longer and more detailed than usual but it is felt that in this case, where the origin and development of a theory is being followed rather than a collection of otherwise related works being described, the length is justified. The methods used are also critically evaluated at some length as the author is of the opinion that sometimes insufficient attention is given to the reliability and precision of palaeoecological techniques and that whereas a short paper may not contain such details, their inclusion in a monograph is thought to be pertinent.

The definition and nomenclature of peat-forming plant communities have been reviewed by Tansley (1939), Du Rietz (1950), Gorham (1957), McVean & Ratcliffe (1962) and Moore & Bellamy (1974). While it may be argued that English terms such as 'raised bog' and 'fen' should be replaced by the more neutral term 'mire', qualified by a prefix such as 'ombrotrophic', 'topogenous' or 'minerotrophic', the common use of the word 'bog' to mean the acidic end of the mire range is too well-established to be ignored. Furthermore, the Norse-derived term 'moss', meaning a bog as well as a bryophyte, is so commonly used in north-west England (as is 'flow', meaning a wet bog, possibly one which has burst in the past) that no strict separation of usage is maintained in this work, as long as the sense is clear.

Plant nomenclature follows normal practice in using Clapham, Tutin & Warburg (1962) for vascular plants and Richards & Wallace (1950) for bryophytes.

2. PREVIOUS RESEARCH

2.1 IDEAS ON BOG GROWTH PRIOR TO 1910

There are two main sources of information on the ideas and observations on bogs put forward by literary travellers, agriculturalists and naturalists prior to von Post & Sernander's statement in 1910. These are the work of Clements (1916) and of Gorham (1953). The latter paper has been drawn on particularly in this section but both of them are notable in quoting *in extenso* passages from early works which leave no doubt that most of the 'discoveries' of the late 19th and early 20th centuries had been foreshadowed by some quite remarkably accurate observations.

Clements (1916), in his review of early investigations into plant succession cites 14 researchers at some length. Of these no fewer than eight were concerned almost wholly or exclusively with peatland successions, as in his first citation, King (1865) 'On the bogs and loughs of Ireland'. Under most authors given simply a mention in 'other investigations' Clements lists 12 'important monographs' prior to 1865. The reason for all this early interest in bogs is not hard to account for. Since at least Neolithic times man has been interested in this unique plant formation which builds up on its decayed remains. Prehistoric 'interest' probably reached its height in Iron Age times when, at the opening of the cooler, wetter Sub-Atlantic period (circa 500 BC) widespread flooding of bog surfaces occurred all over Western Europe; sealing in the tracks or corduroy roads of Bronze Age times with a layer of fresh unhumified *Sphagnum* peat (Coles *et al* 1975, 1976) and making travel and the tapping of natural resources extremely difficult. This deterioration must have been particularly noticeable in Denmark and North Germany, where there was a high percentage of peatland and to judge from the amount of votive material found in these bogs it is quite plain that a degree of 'bog-worship' existed – no fewer than 690 bodies have been recovered from the bogs, the great majority from Jutland, and many were undoubtedly sacrificial offerings (Glob 1969).

The earliest English writings on bogs recorded by Gorham (1953) are those of Leland, in his Itinerary in England and Wales, written between 1535 and 1543, but his comments are not of great ecological interest, unlike the views of Boate (1652) and King (1685) who comment extensively on the different types of bog, their origin and their growth. However it is not until the beginning of the 19th century that a more exact, scientific approach began to be used, sparked off, according to Gorham, by the need for agricultural expansion which led to attempts at drainage and reclamation of peat lands, including Bolton Fell Moss itself. Gorham speaks of the 'brilliant series of personal observations' of the Scotsman William Aiton (1811). Amongst these observations was the recognition, almost 100 years before its independent restatement by von Post & Sernander (1910), of the hummock-hollow complex and its possible role in the growth of bogs. Because of its interest as the primary statement on this phenomenon and because I can find no reference to it in any of the later literature, save in another paper by Gorham (1957), the relevant passage is quoted below:

> These diversities in the patches or lumps of different kinds of moss in a section, frequently appear above and below each other, as well as on either side. This must proceed from the rapidity of growth and slowness of decay of these plants which grow in the gutters, compared with those which grow on the dry parts of the moss. The Sphagnum, &c. which grow in the stanks, rise speedily in close cushions; and as putrefaction advances slowly on these plants, they soon raise the stank or gutter into a height, and the cushions of the Bryum Hypnoides,* &c. rise like small cocks of hay still higher. The plants that grow on the driest parts of the moss, are not so bulky, do not grow in such clusters, and are more completely reduced by putrefaction after their vegetable life terminates. The consequences are, that in the course of time, the stank or gutter becomes the greatest height, the water takes a new course over what was formerly a dry place, and converts it into a stank, the richer moss plants are banished from that spot, the Sphagnum rises in their stead, and forms a patch or lump of white soft light moss, over a stratum of that which is blacker, more weighty, and more valuable. As often as a change is effected on the humidity of any part of the surface, so often will different herbage be introduced, and the mosses formed from such diversified herbage, will also be different from each other.

Together with another Scotsman, Rennie (1807, 1810) Aiton recognised several other features of bog development including the general plant succession from lakes to fens and bogs, the importance of topography and the chemical conditions promoting peat formation.

Towards the end of the 19th century and in the early years of the 20th, the number of studies relating in some way to bog growth increased greatly. Ganong (1897), in his pioneer work on Canadian raised bogs, draws especially on this new European work (e.g. Früh *et al* 1891, Sernander & Kjellmark 1895) in a paper which includes maps and levelled transects of the bogs investigated and water level and temperature observations. However, Ganong was, like most other authors of the period (see in Clements 1916, 1928), vague as to the rate and the details of bog growth. He hinted at the possibility of cyclic regeneration by noting the presence of lichens in some hollows and speculating as to whether they are there because the conditions 'please them better, or whether

* *Bryum hypnoides* was an early name for *Rhacomitrium lanuginosum.*

their presence in certain places had hindered the growth of the moss, thus making the hollows' (page 139).

The period around the turn of the century saw great advances in the study of peatlands, due primarily to the efforts of German and Scandinavian workers. Blytt published his division of post-glacial time into climatic periods in 1876 using terms which we still find useful – Boreal, Atlantic, etc. A vigorous debate followed which did much to clarify ideas on climatic and bog succession, Andersson opposing Blytt's divisions and seeing only one optimum of post-glacial climate followed by a decline (Andersson 1910, and in various papers quoted by Clements 1916) while Sernander accepted the new division, modifying them somewhat in a series of papers culminating in 1908. The divisions are hence known as the Blytt-Sernander periods (West 1968). Some of Blytt's evidence for dry periods was the existence of layers of tree stumps in peat bogs, associated with highly-humified black peat. In 1900 the German investigator C.A. Weber coined the term *Grenzhorizont* (boundary horizon) for the usually sharp division, within the uppermost peat, between the lower dark, highly-humified *Sphagnum-Calluna-Eriophorum* peat and the upper pale fresh *Sphagnum* peat. The date of this *Grenzhorizont* was estimated as Late Bronze Age-Iron Age, about 800-500 BC, on archaeological grounds, though Weber originally thought that a hiatus of up to 1 000 years could have supervened between the two peat layers, due to the late Sub-Boreal warmth or dryness postulated by Blytt and Sernander, causing cessation of bog growth, oxidation of the peat and birch and pine colonisation. Then the regrowth of *Sphagnum* peat began due to a sudden deterioration to cool and wet summers, and within this fresh *Sphagnum* peat Weber described 'a kind of lenticular structure' (Tolonen 1971) without developing this into a theory of cyclic alternation of hummocks and hollows.

Weber followed up this paper with two further ones in 1902 and 1908, again concerned with the development of North German bogs, and putting forward his 'terrestrialisation' hypothesis *(Verlandunghypothese),* all of which laid the ground for many later investigations (e.g. Granlund 1932, Kulczynski 1939, up to Nilsson 1964 and Dickinson 1975 – see later sections). Weber was also one of the first to use pollen analysis on a quantitative basis (Faegri & Iversen 1975) and can thus be looked upon as one of the founders of Quaternary palaeoecology.

2.2 FORMULATION OF THE CYCLIC REGENERATION THEORY

As Koestler (1969, Ch. X) demonstrates no scientific theory arises entirely from a void. The two publications to be discussed in this section, von Post & Sernander (1910) and Osvald (1923), clearly follow on lines of thought deve-

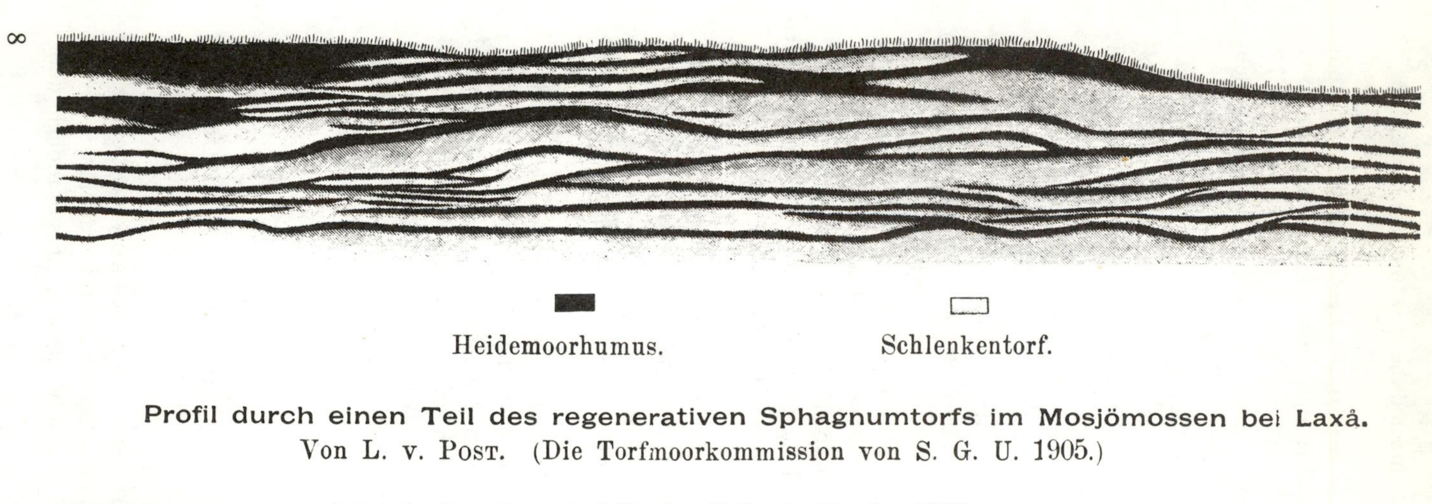

Figure 2. Regenerative peat stratigraphy (from Sernander & Von Post 1910, after Von Post 1905).

loped earlier by such as Weber, and, although they are not mentioned in the above studies, the ideas of evolution and cyclicity current amongst biologists such as Darwin and geographers such as Davis, permeating the scientific world at the time, must have had an influence.

Both these papers are in German and have not been translated in their entirety by the author, but a professional translation of the key chapter of Osvald's monograph was commissioned. Von Post & Sernander (1910) was translated in part and compared with the accounts of it in Osvald (1923), Kulczynski (1949) and other papers. Attention was also paid to other papers by Sernander (1910a, 1910b); important statements from these have been translated and appear in Clements (1916, 1928). In any case, Osvald (1923, p.275) observes that von Post & Sernander's picture of the progress of the succession is *'sehr verwickelter'* – very entangled or complicated. This he ascribed to the fact that they did not have the plant 'complex-concept/idea' *(Komplex-begriff)*, that is, the idea of different species occurring together as indicative of certain environmental conditions or stages of development.

This lack of a firm basis in the ecology and associations of the species concerned in the 'regeneration' is shown in the following excerpts from papers by Sernander (1910a, b), quoted by Clements (1916, 1928) in the course of an involved argument over the impossibility of 'regressive' development:

Sernander (1910:208) has drawn a distinction between progressive and regenerative development:

'The real cause why the *Sphagnum* peat is heaped up in such fashion lies in the fact that the moribund parts lag behind the living *Sphagnum* in growth, and finally form hollows in the latter. These hollows fill gradually with water, while the erosion of the surrounding peat-walls increases their extent. In the water arise new *Sphagneta*, which begin in miniature the progressive development which I term *regeneration*. This regenerative development of the hollows soon culminates in *Calluna*-heath or is interrupted by a new formation of hollows. The latter develops in the usual way, and in this manner arises one lens-shaped peat-mass above another, characterized above and below by dark streaks, usually of heath-peat.'

In discussing the origin of the high moor of Örsmossen (1910:1296) Sernander states that:

'After the progressive development, where regeneration plays a relatively minor role, appears a stage in which the moor passes simultaneously into heath-moor over large areas with uniform topography. (In the deeper hollows, the progressive development may proceed further). In the sequence of the layers, the lower *Sphagnum* peat is followed by a more or less coherent layer of heath peat. With the development of the heath moor begins the formation of hollows, and the accumulation of regenerative peat masses, commonly with great sods of *Andromeda-Sphagnum* peat and *Scheuchzeria-Sphagnum* peat directly above the peat of the heath moor.'

Sernander's description of the formation of hollows by the death of the peat and of the consequent production of tiny pools which are invaded by *Sphagnum* furnishes outstanding proof that the retrogressive development of Nilsson and Cajander is actually the death of a plant community or a part of it, and the resulting formation of a bare area for colonization. No serious objection can be brought against the use of the term *regeneration*

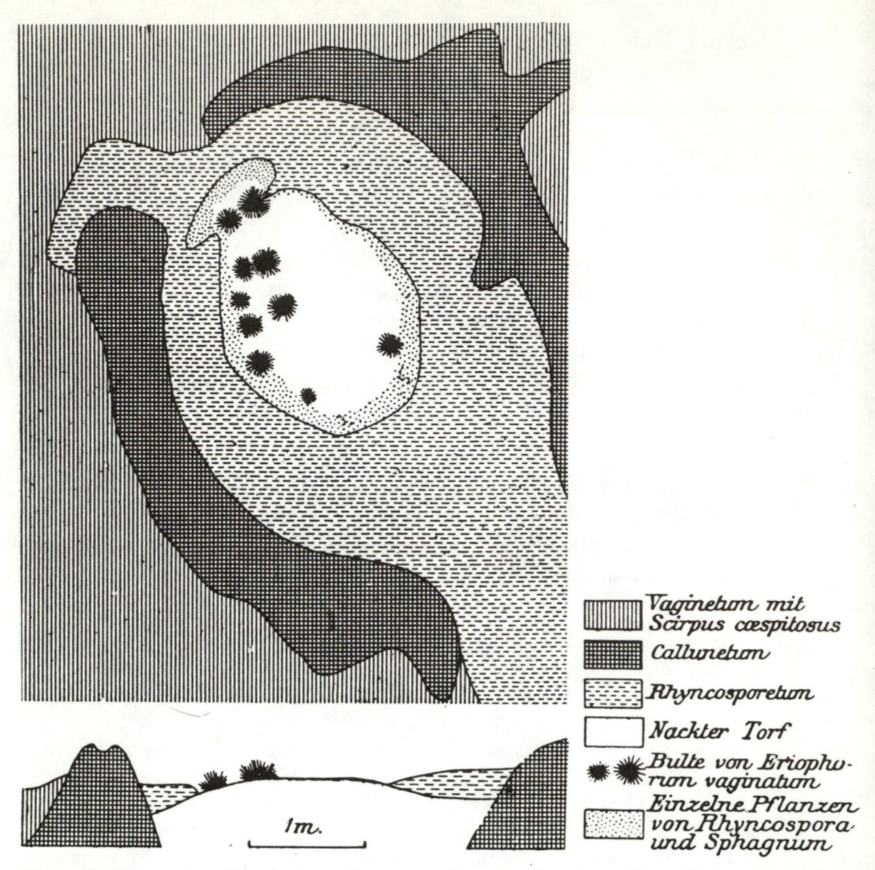

Figure 3. Plan and sectional views of regenerative peat situations (from Sernander & Von Post 1910).

> or *regenerative development*, and it has the advantage of being in harmony with the idea that succession is a reproductive process. It does, however, obscure the fact that the development is nothing but the normal progressive movement typical of succession. It is normally secondary, but differs from the primary progressive development only in being shorter and in occurring in miniature in hundreds of tiny areas. (From Clements 1928).

To Osvald's observations can be added the comment of Tolonen (1971) that 'the stratigraphical material presented (by von Post & Sernander 1910) in support of the theory was rather scanty and inconclusive' and this verdict is borne out when one looks at their diagrams, reproduced as figures 2 and 3.

The single stratigraphic section is generalized to consist of only two peat types, broadly *Calluna*-peat and hollow peat. It was, of course, only meant to convey a general impression of the field stratigraphy but it was upon this stratigraphic evidence, and that presented in figure 3 (and two other drawings very similar to figure 3), that Osvald (1923) built and refined the cyclic theory.

It is interesting to compare figure 1 with plate I, which is of very similar stratigraphy photographed by the author during a visit to Kilberry Bog in Ireland in 1968. The similarity is in parts remarkable but at Kilberry Bog the *Calluna*-rich 'streaks' may be seen to be growing outwards from two distinct hummock areas and, with Walker (1961), the author believes such stratigraphy to reflect the local effects of slightly wetter and drier years, or more likely runs of a number of years, allowing a temporary spread of hummock species across an intervening hollow. Naturally, detailed analyses would be needed to establish this conclusively but it may be noted that the depth of peat section exposed in plate I probably took at least 3,000 years to accumulate (at the average rate of 1 cm every 20 years quoted by Walker (1961)), during which period of time climatic fluctuations are bound to have occurred. No scale is given on von Post & Sernander's diagram but a similar sort of time-scale, and origin, may well be conjectured. In three other important papers on this topic (Walker 1961, Walker & Walker 1961, Hansen 1966) von Post & Sernander are not mentioned, and Casparie, both in his paper on hummock and hollow formation (1969) and in his comprehensive monography on bog development in south-eastern Drenthe (1972) does not mention the original joint paper of 1910 at all and only in passing does he refer to Sernander's role in formulating the theory.

It is clear from this, and from my own study of von Post & Sernander (1910) that although they must be credited with the idea of cyclic alternation it was Osvald's work, particularly his 1923 and 1949 monographs, which established the theory. It should be noted that the 1949 publication was based on fieldwork carried out in 1921, 1931, 1935 and 1937, written up just prior to the Second World War, then delayed and finally published without revision for the 1949 International Phytogeographical Excursion to Ireland. As Osvald comments in his foreword, 'now or never seemed to be the alternatives'. It was therefore written without reference to Kulczynski's monograph (Polish edition 1939 but English translation 1949) or, for example, Overbeck (1946).

Osvald's 1923 monograph is a thorough work of some 436 pages. Chapters I and II are an introduction to the area of the Komosse, near Jönköping in southern Sweden, and the climate; chapter III deals with methodology and then with the plant associations (164 in all) and their groupings, while chapter IV is concerned with the area covered by each association. Chapter V is the core of the work (*Die Assoziationkomplexe,* pp.266-305), which I have had translated in its entirety. Chapter VI is composed of studies on the individual bogs (over a dozen) while chapter VII places the Komosse bogs in their regional and world setting, based on both the literature of the time and Osvald's field visits to, for example, Foulshaw and Deer Dyke Mosses on the northern shore of Morecambe Bay. It should be noted that Osvald uses the phytosociological terms 'association' and 'association-complex' in the sense of the Uppsala School, led principally by Du Reitz (1921) and not in the sense of the Zürich-Montpellier School, led by Braun-Blanquet (1932). The two schools of thought on vegetation classification are now more or less merged, coming together particularly since Du Reitz (1930), but the details need not concern us here except to note that, as Osvald was working within the older tradition his 'associations' refer not to abstract units synthesized from plant lists but were looked upon as real or 'concrete' units to be seen and analysed in the field (Shimwell 1971). His 'association-complexes' were the equivalent of Braun-Blanquet's 'associations', that is, constructed afterwards by inspection of the plant lists, but still recognizable in the field.

Osvald's chapter V has eleven sections describing the different association-complexes and some of the relationships between them. All the associations have been assessed quantitatively by line-transects, and subjected to some statistical analysis. The sections are as follows:

(From Osvald 1923, p.xvi)

As can be seen from the above the *'Regenerationskomplex (Haupttypus)'* is is the most important covering 12 pages; Osvald specifically states this and also that it covers the greatest area on the two main bogs of the Komosse – Slattmossen and Timmerhultsmossen. He also defines it quite clearly (p.268): 'It is mostly built up of associations with a vigorous vertical growth, and as regeneration occurs normally and without hindrance, I called it Regeneration Complex to differentiate it from others, less important, where regeneration occurs as well, I called it *Haupttypus* (chief-type)'. This particular use of the world 'regeneration' is clearly not meant to imply the existence of a cycle but is used in Sernander's (1910) sense of 'progressive development' and later in the chapter there is also no simple equation of regeneration-complex = cyclic alternation. Rather the word 'regeneration' is used as it is defined in Concise Oxford Dictionary (1976) – 'breathe new and more vigorous . . . life into, . . . generate again, form afresh' and the synonyms given in Roget's Thesaurus (1962) give similar meanings: 'rejuvenate, revitalize, revivify', etc. I mention this particularly because some confusion has arisen in the literature over this word; caused in part by Osvald himself. In his 1949 paper he states (p.10): 'The course of this cycle is called 'regeneration' and the resulting complex community regeneration complex'. In later papers such as Walker (1961), Walker & Walker (1961), Casparie (1972) and others, 'regeneration' is used to mean 'growth', 'rejuvenation' and 'hummock-hollow cycle'. One is reminded of Clements' (1928) comment on the introduction of the term by Sernander – 'No serious objection can be brought against the use of the term *regeneration* or *regenerative development,* and it has the advantage of being in harmony with the idea that succession is a reproductive process. It does, however, obscure the fact that the development is nothing but the normal progressive movement typical of succession'.

Were all this and other ramifications (e.g. Watt (1947) on 'Pattern and Process'), to have been foreseen by Osvald in 1923 it is possible that he might have used a more neutral term, such as 'hummock-hollow complex', but as it is 'regeneration complex' cannot be erased from the literature and must be commonly regarded as synonymous with 'hummock-hollow complex', whether cyclic or not.

Osvald's (1923) regeneration complex is made up of the following plant communities, as well as open water and bare peat:

Table 5: Die Zusammensetzung des Regenerationskomplexes

	Timmerhultsmossen		Slättmossen	
Cladonia rangiferina-Ass.			0,04	
Grimmia hypnoides-Ass.			0,01	
Calluna-Cladonia-Ass.	5,56		10,32	
Empetrum-Cladonia-Ass.			0,15	
Erica-Cladonia-Ass.	0,07		0,06	
Calluna-Hylocomium-Ass.		5,63	0,05	10,63
Calluna-Sphagnum magellanicum-Ass.	63,48		47,25	
Calluna-Sphagnum tenellum-Ass.		63,48	1,34	48,59
Nackte Rhynchospora alba-Ass.			0,02	
Eriophorum vaginatum-Sph. magellanicum-Ass.	13,55		14,93	
Rhynchospora alba-Sphagnum tenellum-Ass.			0,47	
Scirpus austriacus-Sphagnum papillosum-Ass.		13,55	0,25	15,67
Jungermania-Schlenken	0,10			
Sphagnum Schlenken	1,04		2,20	
Zygogonium-Schlenken	0,80		3,94	
Sphagnum-Jungermania-Schlenken	2,96		1,47	
Sphagnum-Zygogonium-Schlenken	3,06		7,96	
Sphagnum-Jungermania-Zygogonium-Schlenken	9,25	17,21	8,65	24,22
Teiche	0,13	0,13	0,88	0,88
Erodierte Torflächen			0,01	0,01
Summe		100,00		100,00

(From Osvald 1923, p.269)

As can be seen from this table only three associations are of importance in percentage cover terms, besides the various hollow associations – *Calluna-Cladonia*-Ass., *Calluna-Sphagnum magellanicum*-Ass. and *Eriophorum vaginatum-S. magellanicum*-Ass. These are the same associations as dominate the present surface of Bolton Fell Moss, though there are a few minor differences in composition such as the rarity of *Sphagnum fuscum* at the Cumbrian site, its role being largely taken over by *Sphagnum rubellum*. This similarity in vegetation was felt to be a most important factor in this study, allowing more or less direct comparisons of, for example, hummock-building successions.

The species succession and the build-up of peat is described in a number of places in Osvald's text and since the original is in German two key extracts are given below in translation:

> The lichen-rich heath cannot grow and as the surrounding *Calluna*-moor becomes higher the heath becomes a pit (Grube) and so can cause the formation of a hollow (Schlenken).
>
> The formation of hummocks and hollows can be traced back to the irregular growth of the *Sphagnum* cover. The Sphagna cannot thrive on the higher stretches, they die off and are replaced by non-peat forming communities which drown at the next stage because of the growth of the neighbouring parts.
> (Page 275)

And again, on page 276, the succession is described, with more details of the species present:

> In the damp hollows the *Sphagnum* carpet is often very light, but it becomes slowly thicker and forms the nearly pure *Sphagnum* hollow. *Eriophorum vaginatum* immigrates into these communities. In those parts of the hollows where Sphagna are absent *Eriophorum* is rarer as a strong immigrant. The few that are found are usually relics of the past . . . The hollows, the *Eriophorum vaginatum-S. magellanicum*-Ass. and the *Calluna-Cladonia rangiferina*-Ass. form the foundation on which a series of associations occur as insignificant additions. Because of the differences in the peat which is formed by the four communities the profile obtained in this complex has a typical 'linsenstruktur' (lens structure). If this structure is less prominent one can assume that the heath-community was only of short duration.
>
> Occasionally one gets an interesting addition to this series of successions because the heath-communities regenerate themselves. Usually it is *Sphagnum imbricatum* which starts this off and causes a strong vertical growth in a very small area, which is sometimes as small as 1 m^2. After it has grown high enough it is killed by *Cladonia* and so we get an *Empetrum*-heath which is rich in lichens.

This succession of the typical 'regeneration-complex' is shown schematically in figure 4 (Osvald's figure 54). It is noticeable that there is no clear 'spelling-out' of the idea of cyclic alternation here, nor is there in the whole of chapter V, nor, as far as I have been able to discern, at any other place in the whole monograph. Rather the cycle is implied in, for example, the succession diagram and in certain phrases in the above extracts. It is also noticeable that no consideration is given in Osvald's monograph of the rates of growth of the individual associations nor to the rates of peat accumulation, except in the most general

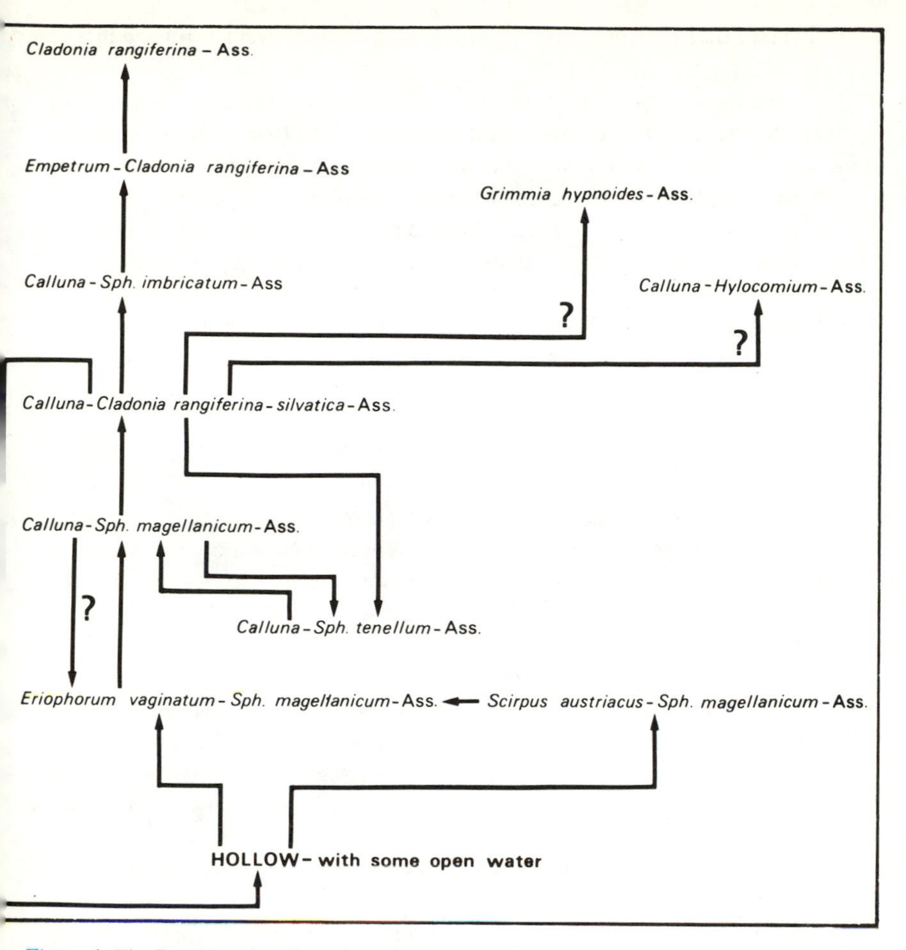

Figure 4. The Regeneration Complex succession on the Komosse bogs (from Osvald 1923).

terms. One must, of course, bear in mind that this pioneer work was primarily one of plant sociology, of the description of communities and their interrelationships, and it can be argued with some justification that speculation on peat stratigraphy was outside the scope of the study. No pollen analyses were undertaken, although levelled soundings (without stratigraphical detail) of all the bogs are given as end-papers, and, of course, all Osvald's work was undertaken before accurate dating via radiocarbon assay was available.

However it is perhaps surprising that nowhere in the main works of Osvald is there any reference to detailed peat stratigraphy or to the possible effects of climatic change. It is true that he was aware of work such as that of Granlund

(1932) and that together with Tansley and Godwin (Osvald 1949) he interpreted the stratigraphy of some peat borings (*not* sections) in terms of the hummock-hollow cycle, but it is clear that Osvald was satisfied with the stratigraphic evidence provided by von Post & Sernander (1910) and looked upon his role as a phytosociologist as merely confirming their ideas, and the earlier observations of Weber, and giving the plant sociological 'mechanism' whereby such successions could take place. His evidence is therefore by no means conclusive and contains a number of assumptions, especially regarding climate and the growth rate of plant associations and of peat.

In dealing with the other complexes of the Komosse bogs Osvald mentions 'regeneration' on several occasions, using the term in a non-cyclical sense simply to mean the build-up of peat and succession of plant communities. For example, the infilling of large ponds in the *Teichkomplex* (p.285) ponds which seem to be in the main residual masses of water left from the lake infill process, proceeds via either *Carex limosa-Sphagnum cuspidatum*-Ass. or *Scheuchzeria-S. cuspidatum*-Ass. to the *Eriophorum vaginatum-S. magellanicum*-Ass. stage of the normal Regeneration Complex. There is no mention of any possibility of a return to open ponds. In the *Randkomplex* there is 'strong vertical growth' and *'linsenstruktur'* can also be found in the peat (p.289) but it is not 'typical regeneration' being mainly an alternation of *Calluna* and *Eriophorum* communities. The culmination of the upward growth of the *Regenerationskomplex* is the *Calluna-Cladonia* dominated *'Stillstandcomplex'* or 'stagnation complex'. No regeneration occurs in this – i.e. there is no peat formed – but there is also little erosion. When erosion supervenes we pass into another non-regenerative, indeed, degenerative complex, the *Erosionskomplex*. We see again then, that 'regeneration' is a somewhat elastic term, loosely used to mean 'strong upward growth' but *'in ihrer typischen Form'* it refers only to the *Regenerationskomplex (Haupttypus)*.

Osvald's 1923 study has been dealt with at some length because of its formative influence on later workers, all of whom cite it, though usually without any detail of its contents. Whilst it must be acknowledged as one of the great achievements of the early Uppsala School it must be remembered that it does not contain any details of plant or peat growth rates, nor any peat stratigraphy, nor any real consideration of recent climatic change and its effects on the plant communities of ombrotrophic mires. And although Osvald obviously accepted the basic idea of the Regeneration Complex as a cyclic succession in his later work (Osvald 1949) and discussed this on his visits to Ireland with Tansley and Godwin (Tansley 1939) and to Wales with Godwin, and Conway (1937 visit; Godwin & Conway 1939), there is no *explicit* discussion of the cycle in this early monograph.

2.3 CONFIRMATORY EVIDENCE AND EARLY DOUBTS

The studies considered in this section are a selection of the literature which appeared on bogs up to the beginning of the Second World War. This period is not as arbitrary as it may at first appear. It saw a number of studies which agreed with Osvald's theory, culminating in Tansley's 'The British Islands and their Vegetation' (1939), and a few which expressed tentative doubts as to the reality of cyclic regeneration, though broadly accepting it (Godwin & Conway 1939). It also saw the completion of two major studies on bog ecology which were not, however, published until after the war as have already been mentioned – Osvald's 1949 monograph, the manuscript completed 'just before the war', and Kulczynski's study *'Torfowiska Polesia'*, Polish edition 1939, English translation 1949. This must be borne in mind when considering the state of knowledge of authors at various dates.

The further works of Osvald (1925a, 1925b, 1933 and 1937) add little real substance to the theory. They are again almost purely works of phytosociology and describe the plant associations of areas other than the Komosse. There is constant referral back to the Komosse situation and the different species encountered are equated with Komosse-equvalents. For example, in his 1925a paper *Sphagnum fuscum* is noted as the main hummock builder in the 'eastern facies' of regeneration complex bog (Eastern Sweden, Finland, etc.), in the place of the more oceanic species, *S. magellanicum.* These papers, besides their intrinsic importance as detailed descriptions of bog vegetation, also served to spread Osvald's ideas, on regeneration in particular, to a wider audience. Taken with his 1949 monograph we get three major works in German, one in Swedish and two in English, a spread of language unrivalled by other workers (e.g. Weber, Overbeck), and which must have been a factor in the dissemination of his views. One may speculate on how many workers would know of Kulczynski's work had it remained in its original Polish.

The first paper to shed any doubt on the occurrence of the hummock-hollow cycle, after the publication of Osvald (1923), was that of Katz (1926). He also used the methods of the Uppsala School (Du Rietz 1921) to describe Associations and Association-Complexes in the *Sphagnum* bogs of Central Russia but notes that the changes in Association-Complexes are rather different to that which obtains in more northern latitudes. Instead of the hollow communities growing up and overtopping the old hummocks, these hollow communities are very susceptible to summer drought and the Sphagna in them (*S. balticum* mainly rather than *S. cuspidatum*) die off over areas of thousands of square metres. The tussock Sphagna, on the other hand, 'do not die off due to drought, nor are they replaced by liverworts and lichens as in more northerly latitudes' (p.192). If the hollow Sphagna do not survive the drought the hollow is recolonized by *'Hypnum fluitans' (= Drepanocladus fluitans* (Heder.) Warnst;

Richards & Wallace 1950) and *Rhynchospora alba* (cf. the *'Rhynchospora*-rich regeneration complex' of Osvald 1923) and then *Eriophorum vaginatum* which grows up in tussock-form accompanied by *'Sphagnum medium' (= S. magellanicum).* Katz then sees these new hummocks and the old ones 'flatten out the microrelief into an approximately level surface' of *Eriophorum vaginatum*-Sphagna-association. *Calluna vulgaris* and other Ericaceae such as *Erica tetralix* are, of course, absent from the area, both being oceanic species, the latter more so than the former (Gimingham 1960, Bannister 1966). Their places are taken by *Ledum palustre, Cassandra* (now *Chamaedaphne*) *calyculata* and *Pinus sylvestris* which top the bog hummocks. It is debatable whether this difference in hummock-top species would be crucial in deciding whether or not the cycle took place but one cannot imagine it to be a vital factor; *Calluna vulgaris* is rare in the Polesie bogs (Kulczynski 1939/49) but 'lenticular regeneration' was thought to occur, with other species contributing to the hummock-top vegetation and Katz does mention the lichen *Cladonia rangiferina* as a hummock-top species which is surely more important as an agent of 'hummock-degeneration' (Osvald 1923). The role of *Calluna vulgaris* in bogs is in any case different to its role in heathlands, as is discussed later.

As to the eventual fate of the *Eriophorum vaginatum-S. magellanicum* association, Katz merely says that – 'In time differentiation of the microrelief begins afresh and the gradual development of a new complex ensues' – there is no mention of a cyclical 'regeneration complex'. Finally one is struck by the last three pages of Katz's paper, on the influence of cultivation and drainage, extraction of peat, and fires on the bog vegetation. He sees those 'external' factors as having a great influence and postulates various types of succession following such interference, hinting that these have not always been recognized and noting that, for example, *'Eriophorum vaginatum* develops luxuriantl after a moderate fire, a phenomenon which the author has witnessed at Bolton Fell Moss.

In 1932 Granlund published his work on the stratigraphy of Swedish bogs, which included the recognition of four 'recurrence' surfaces (Swedish: *recurrensytor* = RY) besides the *Grenzhorizont* of Weber, to which he gave dates of 2300 BC (RYV), 1200 BC (RYIV), 600 BC (RYIII, the *'Grenz'*), 400 AD (RYII) and 1200 AD (RYI). These recurrence surfaces have been described and studied by many later authors but the point of immediate interest with regard to the 'regeneration complex' is the existence of RYI at about 1200 AD. Quite obviously a widespread drying out of bog surfaces at that date, represented by peat at a depth of only 20-50 cm (depending on the peat accumulation rate since 1200 AD), must have a bearing on the hydrology and vegetation of the present bog surface and the thought also presents itself that if a fairly substantial climatic change has the effect observed as a recurrence surface, could not less marked climatic changes effect the bog vegetation in less noticeable but still important ways?

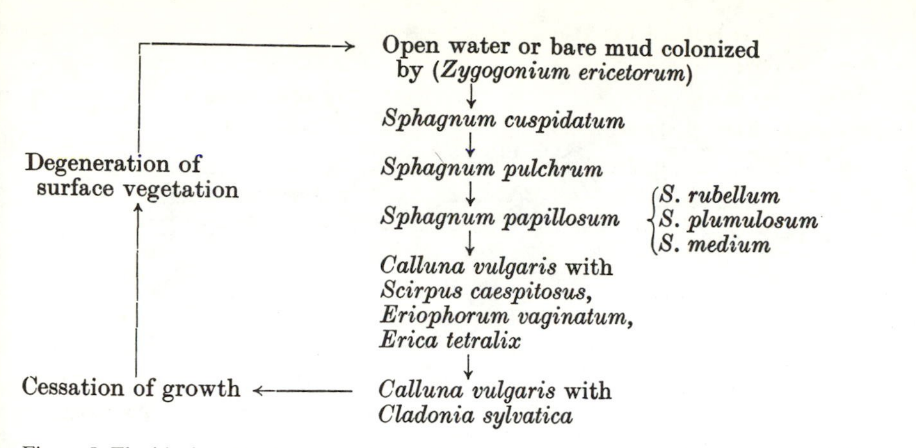

Figure 5. The ideal cycle of regeneration at Tregaron Bog (from Godwin & Conway (1939).

This sort of speculation does not appear to have been in the mind of, for example, Ernst (1934) who saw hummocks and hollows as an autogenic cylic system following Osvald (1923), in the bogs he studied in North Friesland, nor is it given any prominence in Osvald's or Kulczynski's work in the 1930's.

The work of Godwin & Conway (1939) on Tregaron bog did much to establish Osvald's theory of regeneration in the English speaking world. Osvald visited the site for a day and it is tempting to see his influence in the following description:

> The best defined and most typical facies of the community (Sphagnetum) is the Regeneration Complex.
>
> The mosaic of plant associations or subassociations which cover the ground here can only be interpreted in terms of local cycles of succession in which *Sphagnum* species colonize the bare mud of pools, and grow rapidly to build up the surface till it is dry enough to bear *Calluna,* thus preventing further growth of *Sphagnum.* Lichens, especially *Cladonia,* come in at this stage, which is followed by a breakdown of the raised tussock, leading to the re-formation of a depressed area, so that a pool appears in the middle of the tussocks that have meanwhile been building up around it.

The 'ideal cycle' at Tregaron is then given as shown in Figure 5. Godwin & Conway comment that this cycle 'cannot be demonstrated clearly in more than a few spots, the degradation stages being particularly hard to find' and give no further evidence or comment on the reality of the cycle until the last four pages of the paper when some interesting comments on climatic change and stratigraphy are made. Figure 5a, which is figure 34, p.357 of Godwin & Conway (1939), shows the results of 'careful excavation in the Regeneration Complex'. Shallow sections in the Molinietum and Scirpetum (Trichophoretum) gave the same result – a sharp discontinuity just below the present surface and

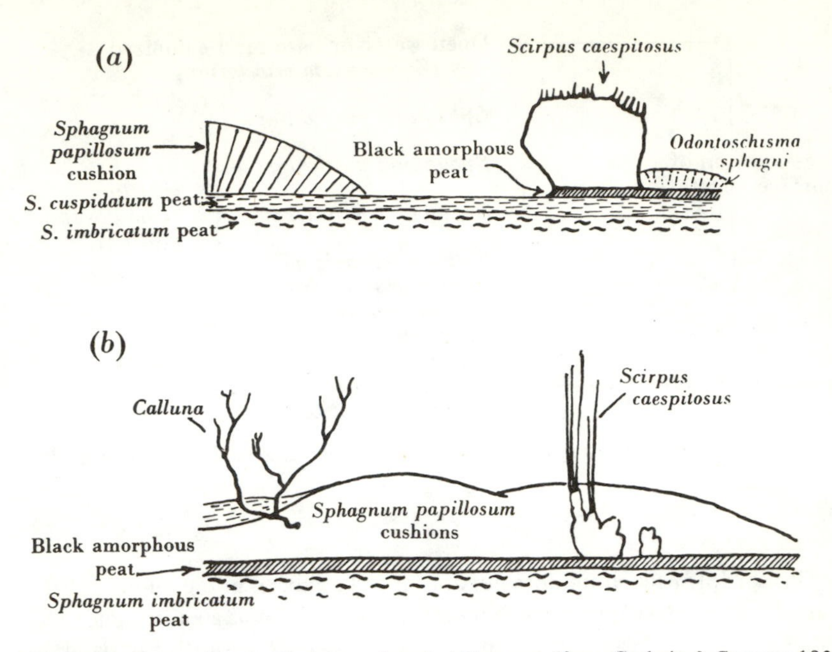

Figure 5a. Excavations in the topmost peat at Tregaron (from Godwin & Conway 1939). Diagrams of soil profiles showing transition from underlying *Sphagnum imbricatum* peat to the most recent peat immediately below the existing vegetation.

then moderately fresh *Sphagnum imbricatum* peat. As this discontinuity extends under all the main bog associations and there is sometimes a layer of highly humified peat just under the present surface they conclude that something analogous to a recurrence surface is presently forming at Tregaron, that is, it is due to climatic change. They concede that the same effect may have been caused by drainage or by burning, dismissing the former factor adequatel but not discussing the possible influence of fires. They conclude the paper with an illuminating paragraph:

> It is interesting, in the light of our present interpretation of recent bog history, to recall the comments made in describing the Regeneration Complex (p.327) to the effect that stages of pool invasion were very frequent and degenerative tussocks very infrequent. This departure from expectation is not remarkable if the Regeneration Complex is only just re-establishing itself over an older drier surface. We feel that possibly similar anomalie not only in Tregaron, but in other West European bogs, could be resolved by recognizing the importance of recent climatic fluctuation.

The nature and timing of such climatic change is left unexplored by Godwin & Conway, though of course much of the evidence presented in the present work was not available in the 1930's. However, the immediately sub-surface stratigraphy is strongly suggestive of burning. Although the drawings (figure 5a) are without a scale the black amorphous peat appears to be a few centimetres thick, at most 5 cm, above moderately fresh peat; this contrasts with the *Grenz*-type structure of highly-humified peat *culminating* in an even more humified layer. Also, in their figure 13 (page 336) 'mats' of *Aulacomnium palustre* and *Hypnum scheberi* are shown overlying a series of 'stumps' of *Eriophorum vaginatum* and *Scirpus (Tricophorum) caespitosus.* It is difficult to see how such 'tussock stumps' could arise by drying out or any other agency apart from a fairly intense fire. The moss 'mats', of the species named above, which are not markedly oligotrophic, could then have benefitted from the resulting ash layer. Valuable as Godwin & Conway's paper is as a description of the ecology of Tregaron Bog in the 1930's, and although the existence of a 'regeneration-complex' mosaic is clearly demonstrated, the existence of the cyclical process postulated to occur within the mosaic remains in doubt.

Tansley's great study, 'The British Islands and their Vegetation' (1939), depends very heavily upon Osvald's work and guidance in the field, as well as Godwin & Conway's Tregaron work in which both senior workers took part. This reflects the paucity of work on British bog ecology up to that time; the broad-scale surveys of Lewis (1904), Rankin (1911) and Moss (1911) were not really comparable with Osvald's Komosse work, Pearsall (1938) had only just begun to make observations on pH and 'redox potentials' and there was no systematic Osvald-style description of a British raised bog. Osvald's views are accepted without demur as the footnote on page 688 makes clear:

In the summer of 1935 the author had the privilege of examining some Irish raised bogs in the company of Prof. Osvald, who interpreted the details of their vegetation and peat structure in the field: the present account was thus made possible.

The statement on the 'hollow-hummock cycle' is direct and presented as observable fact rather than as a theory to be tested:

. . . above this (fen peat) come successive cycles of different kinds of bog peat. The existence of these cycles depends upon the fact that the surface of a bog at any given time is not uniform, but consists of alternating hummocks and hollows inhabited by different species. The peat at the bottom of each hollow is built up by the vegetation in the natural process of autogenic succession until it forms a new hummock, the surfaces of the adjacent pre-existing hummocks which have stopped growing thus coming to occupy a lower level than the new hummocks. The old hummocks thus become the sites of new hollows, the old hummock vegetation dies, and is replaced by species characteristic of hollows, these in their turn giving away to new hummock formers. Thus the structure of the peat of a raised bog is lenticular, each lenticle representing a complete cycle of 'hollow-hummock' development, and all the phases of the cycle are represented at any given time on the surface of an actively growing bog ('regeneration complex'). The lenticular structure can be clearly seen in vertical section of the peat. The story of raised bog development was first worked out in detail by Osvald in his classical paper on the Swedish bog 'Komosse', and the Irish raised bogs are essentially similar, though some of the species are different. (From Tansley 1939, p.678-688)

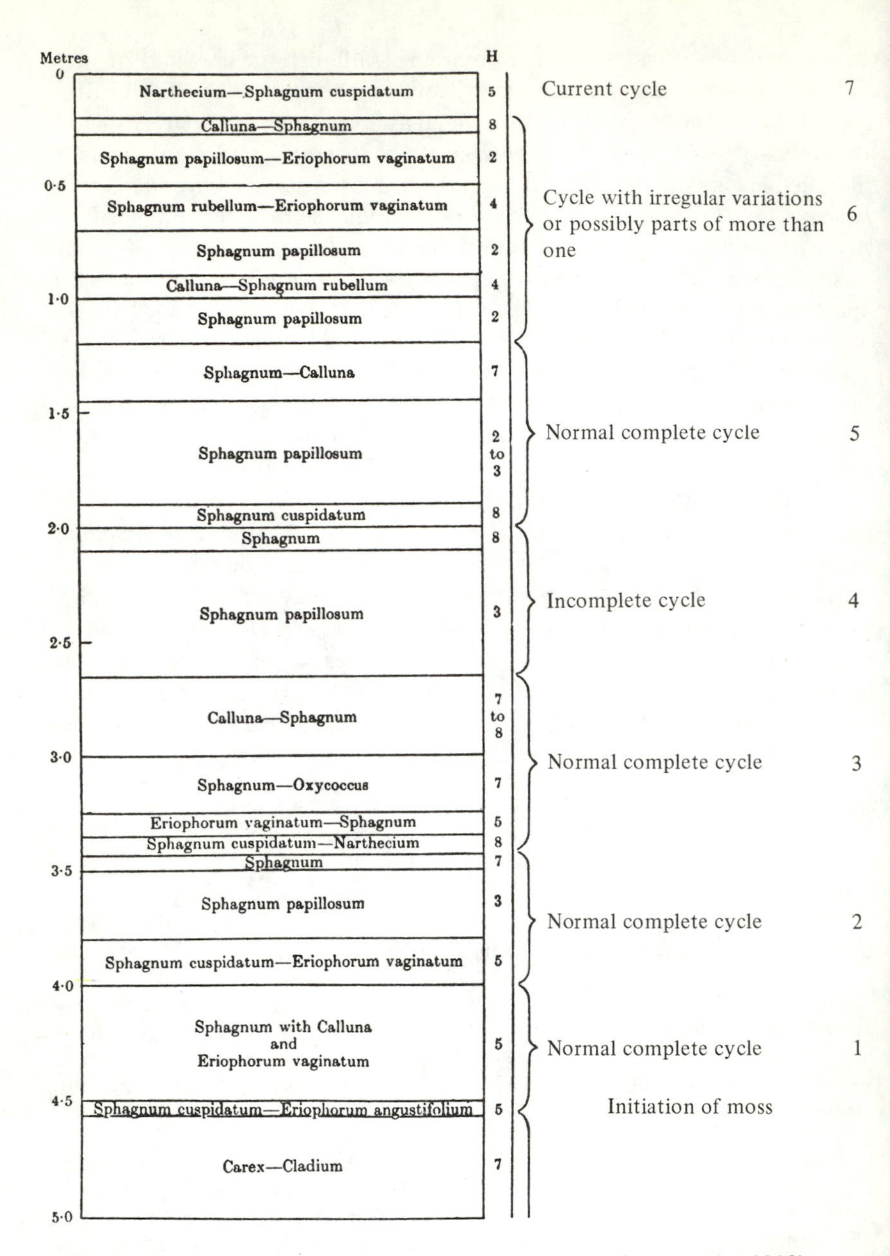

Figure 6. 'Regeneration cycles' constructed from a borehole (from Tansley 1939). Section of bog south-west of Athlone through moss (bog) peat and underlying fen and reedswamp peat to lake marl and basal glacial clay. H = degree of humification (scale of humification 1 to 10).

Plant lists of each stage of the cycle are then given, based on a bog south of Athlone, Eire (no name or location) and some mention is made of the 'structure of the hummocks' based on observations of the surface vegetation. It is noted here that 'on the older hummocks, after the entrance of *Cladonia sylvatica,* the ling tends to become 'leggy' and die'. This cannot, says Tansley, be due to drought and so he invokes insufficient nutrition and the age of the plant.

Tansley then gives (pp.690-696) the results of two borings made by Osvald through the bogs at Athlone and Edenberry and Osvald's interpretation of the ombrogenous peat in terms of 'cycles' and 'incomplete cycles'. It is noted that such a vertical bore will not give a complete picture and may 'pass through the edges of the (assumed) lenticular units' but the bores are nevertheless given prominence as two tables (XXI and XXII) covering four pages. Part of the 'best' of one of these is reproduced as figure 6.

These diagrams give an erroneous visual impression of the borehole – implying by their width sections rather than boreholes – and the claim that these alternations of species represent hollow-hummock cycles, is quite spurious. A 'normal complete cycle' is based simply on the occurrence of a *Sphagnum cuspidatum* band and the possible occurrence of wide swathes of peat of a particular type is ignored (i.e. if *S. cuspidatum* was present it was assumed to represent a hollow, if *Calluna* and *S. papillosum* were encountered it was assumed to be due to hummock building). These ideas could have been tested by examination of the many thousands of peat-cutters sections existing then as now in Irish peat bogs, or else an attempt could have been made to link up two or more closely adjacent boreholes; perhaps this is a classic case of Koestler's 'snow-blindness' (1969, pp.216-220) for in the presence of Osvald, who had done so much classic work 12 years previously, it apparently did not occur to Tansley to use this approach. Similarly, if Tansley ever questioned Osvald on such points as the influence of climatic changes and water-table fluctuations, or on why flat 'lawns' were not produced by the infilling of hollows, the questions and their answers never appeared in Tansley's influential work.

2.4 NEW APPROACHES AND CONTINUING TRADITIONS – 1940-1960

The war years drastically reduced the flow of scientific work in bog ecology but in those papers which did appear in the 1940's and early 1950's, some new approaches were evident. Pearsall (1963) remarked '. . . there is no doubt that each of the great wars we have survived has produced tremendous changes in scientific outlook' and particularly after the Second World War, in which so many scientific advances were made, a rather more comprehensive ecology began to emerge, attempting to consider all the factors affecting a plant community. The seeds of this approach may perhaps be found in Godwin's pre-war work, particularly in pollen analysis (Godwin 1934, 1940) and the fruits began to appear in bog ecology with papers such as Pearsall (1941), Pearsall & Lind (1941) and Harley & Yemm (1942). In these papers there is little mention of the 'regeneration complex' (none of them are classical 'raised bogs' though ombrotrophic to a greater or lesser degree) but there is some speculation and hypothesizing on the origin of the present bog vegetation, and factors such as the possible effects of burning and grazing on different species are discussed. In 1941 and 1946 Godwin published his correlation of bog stratigraphy, hydrology, climate and archaeology in the Somerset Levels. In the later paper he made some interesting observations on recurrence-surfaces, and some criticisms of previous work (Granlund 1932) but still accepted 'Regeneration Complex' peat as the normal deposit of an actively growing raised bog. A little later (Clapham & Godwin 1948) there was published a 'diagrammatic section' of Somerset peat which purports to show lens-shaped masses of 'regeneration complex peat' above a recurrence surface; unfortunately no close-up photographs or detailed drawings of this peat face are given, nor is there any further discussion of the phenomenon.

At the same time Conway (1948), following her pioneering work on Ringinglow Bog, produced a perceptive review of von Post's work on climatic rhythms. In considering 'some scattered pieces of evidence concerning rhythms of still smaller periodicity' she puts forward the following doubts regarding cyclic development and climatic change:

> The regeneration complex vegetation of raised bog has been generally regarded as representing a cyclic type of development which will proceed under stable climatic conditions, if these are of the right sort. Without offering any definite suggestion, one might perhaps at least put forward the question as to whether we should have to alter our interpretation of the observed vegetational mosaic which we call the regeneration complex, if it should be proved that climates are never stable but always in some phase of a larger or smaller oscillation? A relevant paragraph may be quoted without further comment from the account of Tregaron bog given by Godwin & Conway (1939):(From Conway (1948) p.231, and she goes on to quote the paragraph already cited here on page 20.)

Although some of the data quoted by Conway has been shown to be capable of other interpretations and though more exact dating methods have invalidated some of her 'rhythms', her idea of moisture 'thresholds' is still a valuable concept.

The years 1949 and 1950 saw the publication of two monographs, Osvald (1949) and Kulczynski (1949); a new 'popular' book on British vegetation (Tansley 1949), a 1949 'reprint with corrections' of Tansley's earlier work (1939), and the holding of the Seventh International Botanical Congress (1950) in Sweden.

As has already been noted both Osvald's and Kulczynski's works were essentially pre-war though some new material was added. Osvald (1949) reiterates his Komosse-type succession and gives association lists and percentages before his extensive treatment of certain bogs in England and Wales, Ireland and Scotland. Many of these observations were the first systematic descriptions of a number of such sites, and Osvald's visits to the British Isles in the 1920's and 1930's had a great influence on the development of mire ecology here as is obvious from the following remarks from Pearsall (1963):

> My own point of view . . . was in large measure associated with a visit which Hugo Osvald paid to this country in 1921. We travelled round the north of England together looking at bogs and moors and I acquired from him the technique of using metre quadrats as descriptive units. . . Hugo Osvald's visit was I think memorable in many other ways. Professor Godwin will bear me out when I say that Osvald, fresh from his classic work on the great Swedish bog Komosse (1923), made us look at our own peat-moors again and materially altered our methods of approach.

Pearsall goes on to note how it was still (in the 1930's) 'almost an article of faith' that the cotton-grass moors of the Pennines were natural units of vegetation rather than burnt-over and polluted areas, once supporting a varied vegetation including much *Sphagnum* bog. While Osvald certainly did throw new light on such problems (but not alone – see Woodhead 1929 and Erdtman 1929) his 1949 monograph contains no stratigraphic detail of the peat apart from the single boreholes already referred to in Tansley (1939). Peat cuttings were noted (p.29) but no advantage taken of the stratigraphy they displayed. Cyclic succession was taken for granted and a number of succession diagrams, similar to those from the Komosse study (figure 4) were drawn up from observations of the zonation of the surface vegetation. Very little justification or evidence for the retrogressive stage is given, though wind erosion of hummocks is invoked on occasion (p.46). Overall then, although this 1949 monograph certainly provided much new material on bog vegetation and community organisation, it contributes little to the development of the cyclic theory.

Kulczynski's 1949 monograph is a comprehensive survey of some 356 pages, 46 photographs and no fewer than 100 figures – even so it is shorter than the Polish version (1939). It has three main parts, dealing with valley, transition and raised bogs, together with a general section on water movement and bog development. While there are interesting observations and hypotheses in all sections, that section on 'raised bogs', pp.72-168, is the most relevant to the present study, and within this section pages 80-101 deal particularly with the regeneration complex. That the regeneration cycle is accepted by Kulczynski is made clear in his preface (p.6-7):

> I deem it my duty to point out that my contact with foreign scholars of the peat-bog problem in particular the excellent botanical peat school in Sweden, has not been without a decisive influence on the drift and methods of my peat bog investigations in Polesie. The selection of a proper direction in my investigations . . . was the result of becoming personally acquainted with the eminent founder of the Upsala peat school, Prof. R. Sernander, and his distinguished pupils, Dr H. Osvald and Dr E. du Reitz, as well as . . . the magnificent works of von Post and C. Malmström. The reader will notice without difficulty that the theory dealing with continental raised bogs, propounded by me in the present monograph,

constitutes – in spite of certain modifications and differences – nothing more than a further development of the well-known and always fruitful raised-bog theory created by R. Sernander and von Post, and elaborated by H. Osvald.

This final admission is interesting in the light of later workers citations of Kulczynski's work (e.g. Moore & Bellamy 1974) where it is implied that Kulczynski's studies independently support the cyclic theory. It is clear, however, that Kulczynski repeatedly attempts to force diverse observations into a predetermined conceptual mould, that of the 'regeneration cycle'. For example, in describing 'the bog Mak' in Polesie (now known as the Pripet Marshes and no longer in Poland but mainly in Russia) he notes that the surface supported a pine forest until it was deforested by man about 50 years previously. Then '. . . processes of lenticular regeneration set in, developing however only one single succession cycle and at present again arrested'. The only stratigraphical evidence for this lenticular regeneration is in the form of single profiles of less than 50 cm depth and similarly vague evidence is provided for two other bogs. Special cases of single-species regeneration are noted – for example near the artificially waterlogged margin of Moroczno bog 'the plant succession taking place passes thereby through the same three stages which exist in the lenticular regeneration of every separate hummock . . .' but 'they do not form here a complex combined in space but . . . succeed one another in time'. That is to say '. . . instead of lenticular regeneration the bog experiences here only one successive stage at a time on its whole surface', so that even in this case, with a patently non-cyclic succession, Kulczynski attempts a cyclic theory explanation. There is also a lot of

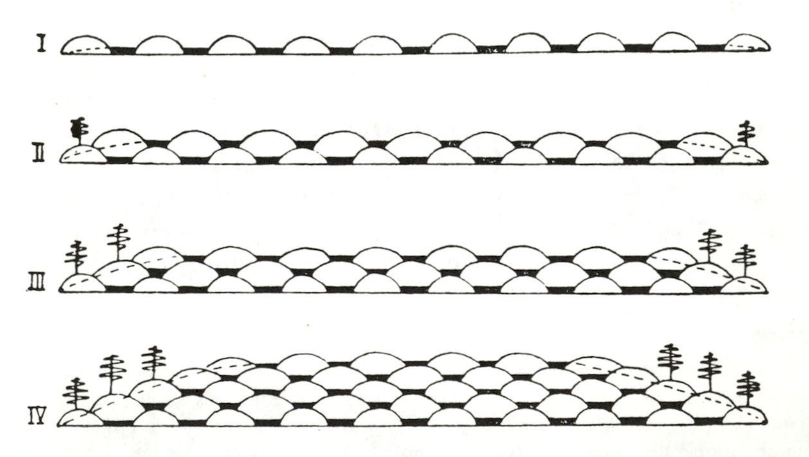

Figure 7. Bog development by lenticular regeneration, according to Kulczynski (1949). Diagram of lenticular regeneration of *Sphagnum* peat. The black colour denotes regeneration hollows (*Sphagnum cuspidatum*), white – regeneration hummocks (*Sphagnum fuscum*). The broken line denotes the water-table in the peat, rising as regeneration proceeds.

comment on the only non-stylised stratigraphic diagram in the monograph, a peat section 2 m long by 1 m deep from a bog near Moscow worked on by Dokturowski. This section (Kulczynski 1949, figure 42, page 121) shows five to six alternating and almost flat layers of *Sphagnum magellanicum* and *S. recurvum* peat. This is ascribed to fluctuating water levels and is not seen to conflict with cyclic regeneration, the latter being the 'normal' mode of bog growth in a stable climate. Kulczynski therefore came close to recognising the importance of climatic variations, but the overall emphasis of the monograph is aptly portrayed by his highly schematic view of bog development (figure 7) which ignores all considerations of scale, and in which there is no place for recurrence surfaces. Finally one must note the following quotation as a clear indication of Kulczynski's views on the cyclic theory:

> The exact course of lenticular regeneration was investigated on its main features by Sernander and von Post. Our knowledge of the whole process was substantially extended by the analytical studies of H. Osvald, so that today the theory of the lenticular regeneration of *Sphagnum* peat stands among the most beautiful and well-grounded achievements of bog science.
> (From Kulczynski 1949).

From what has been discussed already in this section one might question particularly the last phrase of the above quotation, but still recognize the great amount of useful observation and discussion put forward by Kulczynski; his recognition of the hydrological systems at work in valley and transitional bogs, is a feature of particular value in his work.

In 'Britain's Green Mantle' (1949) Tansley propounds the cyclic theory without any reservations, describing the 'striking and very constant zonation (of species) in space' as a 'real succession . . . in time' and claiming that the diagrammatic section reproduced from his earlier work (similar to figure 1) 'is confirmed by vertical borings through the peat'. No mention is made of the *caveats* which appear in the larger work, though in the second edition (1968) the reviser, M.C.F. Proctor, notes at the foot of the page (p.206):

> This cyclic succession is probably less prevalent in actively growing raised bogs than has been supposed. In some cases the stratigraphy of the peat indicates uniform growth of more extensive carpets of *Sphagnum.*

In the four mire excursions associated with the International Botanical Congress of 1950 held in Stockholm, three were led by Du Rietz and one by Osvald. The excursion guides to all of these naturally hold to the cyclic view of bog vegetation, Osvald's account of the Komosse more obviously so than Du Rietz's accounts.

An interesting point to emerge from Osvald's new account of Komosse is his changed emphasis when discussing the area marginal to the Regeneration Complex. On page 5 (Osvald 1950) he describes a walk over a typical bog and notes:

The marginal zone of the bog centre is very smooth and even. *Calluna-Sphagnum magellanicum* (S. fuscum) sociation* and *Calluna-Cladonia* sociation are the dominant plant communities. Hollows and hummocks are very scarce and inconspicuous or totally absent. This is the so-called marginal complex.

It is clear from the map of Slättmossen (figure 2, page 13), which is the same as figure 69, page 318 in Osvald (1923), that this 'marginal complex' is the *'Randkomplex und heidenartiger Randkomplex'* of the older work, wherein the lack of hummocks and hollows is commented on but not stressed as in the above quotation. (Quote from translation of page 289, 1923: '. . . it cannot be said that there are a lot of hummocks'). It is also clear from the levelled profiles (endpaper, Osvald 1923) that the marginal complex is by no means always on the 'rand' slope and Osvald (1950) writes of it: 'Leaving the lagg and the rising margin behind us, we walk out over the surface of the moss; this is an almost horizontal plane with small stunted pines here and there. The marginal zone' . . . (then as in the quotation above). Taking this together with Godwin & Conway's (1939) views, and the views of Godwin (1946) on water levels in bogs and the role of 'retardation layers' in holding up the water-table, the thought presents itself that instead of this marginal complex developing into a regeneration complex (Osvald 1923, p.289: 'During the growth and transgression of the moor it develops into a regeneration complex'), could not the marginal complex represent the stage that the regeneration complex will reach when its pools infill? The following table (Osvald 1950) lends weight to this possibility, for it will be noted that both marginal complexes, but especially the wetter type, are very close in composition to the regeneration complex:

Table 2. The composition of some of the most important complexes on Komosse

Sociation	Marginal complex wetter type %	Marginal complex drier type %	Regeneration complex %	Stagnation complex %	Erosion complex %
Calluna-Cladonia sociation	10	40	10	25	25
Erica-Cladonia sociation	–	–	–	5	10
Calluna-Sphagnum magellanicum sociation	65	50	50	35	10
E.vaginatum-S.magellanicum sociation	20	–	15	–	–
Sphagnum-Gymnocolea hollows	–	–	20	–	–
Zygogonium hollows	–	–	–	30	20
Eroded peat areas	–	–	–	–	25
Others	5	10	5	5	10

(from Osvald 1950)

* Post-1930 nomenclature of Uppsala School. Equivalent of 'Association' of Osvald (1923).

Papers by Sjörs which appeared at about this time (1948, 1950) are quoted by a number of later investigators (e.g. Gorham 1953, Walker 1961) as being relevant to the problem of hummock-hollow alternation. I have been unable to find any direct comment in Sjörs' work on the problem, though he does describe phytosociologically a number of interesting hummock and hollow communities and comments at length on water movements and bog surface patterns. It must be noted though that these patterns are in boreal mires, rather than the sort prevalent in S. Sweden, Germany, Britain, etc., and that frost induced changes are almost certainly of importance in the mires further north.

During the 1950's a number of papers mention bog stratigraphy while dealing with other main themes – for example, vegetational history (Conway 1954) – but none claim to see regeneration cycles in the peat. Others, dealing with recurrence surfaces (e.g. Godwin 1954, Overbeck & Griéz 1954, Olausson 1957) refer to the regeneration complex in passing, either assuming its validity (Godwin 1954) or doubting its occurrence (Olausson 1957), but without microstratigraphic details.

The next paper that deals with the regeneration complex *per se* is by Millington (1954), writing on *Sphagnum* bogs in New South Wales, Australia. These mountain bogs have only one *Sphagnum* species, namely *Sphagnum cristatum* Hampe. I have been unable to trace any data on the ecology of this species though it appears to be related to *Sphagnum centrale* (according to Isoviita 1966) a cymbifolian species 'widely scattered but rare, with a generally northern distribution' (Proctor 1955). It forms a ground layer with the sequence of hummock-building being due to the succession: sedges – restionaceous plants – shrubs. Degeneration of these hummocks occurs by subsidence, water movement changes, shade and wind erosion. Despite these gross differences in vegetation type Millington still recognizes a Regeneration Complex and Stillstand Complex *sensu* Osvald, whom he quotes in justification.

Work rather more easily comparable with Osvald's is that of Morrison (1955, 1959). Special significance attaches to Morrison's work on the Fairy Water Bog, Northern Ireland, as it still has *Sphagnum imbricatum* as the most active hummock former. Morrison contrasts these hummocks with those described by Tansley (1939) and Godwin & Conway (1939) saying that their Regeneration Complex is a 'very poorly marked complex' in comparison and, following Duvigneaud (1949) he regards *Sphagnum imbricatum* and *S. fuscum* as the only 'true hummock' builders. This is, of course, at odds with the originators of the cyclical theory, though Osvald (1923, 1949) recognized occasional secondary hummock formations, building on lower hummocks, as being due to *S. imbricatum.* Perhaps because of the special circumstances prevailing at Fairy Water Bog, within which there is a trapped body of water, a kind of 'Schwingmoor' structure, Morrison sees no distinct cyclic alternation, the pools being 'windows in the crust of the mire'.

In the early 1950's we also see the first modern papers on bog chemistry, papers such as Sjors (1950), Gorham (1949, 1953), Gorham & Pearsall (1956), Mattson & Koulter-Anderson (1954), etc. Though not bearing directly on the cyclic theory they do provide extra evidence of habitat conditions and show that bog ecology was becoming a more 'integrated' affair, a trend exemplified later by Chapman (1964a, b, 1965).

Godwin's 'History of the British Flora' (1956, First edition) a landmark in palaeoecological studies, does not make much of the regeneration complex, saying simply that 'it produces a characteristically banded peat in which the vegetation of the pools and of the hummocks can be traced in constant alternation with one another' (page 30). The same statement appears on page 31, second edition, 1975. Also in the mid-1950's notable papers on bog ecology were published by Pearsall (1956), Gorham (1957) and Overbeck & Happach (1957) but they deal with other points of interest rather than directly on the cyclical theory and its propagation.

In 1958 Ratcliffe and Walker published the results of their thorough investigations on the bogs collectively known as The Silver Flowe, in south-west Scotland. Some of the observations made are of importance in overturning or correcting previous views and amongst these are particularly the results of a survey of species distribution in relation to the water level in pools – the range of all species above and below water level was proved to be rather more extended than was commonly supposed in the literature. Naturally there appeared to be optimal levels for various species but very few species were shown to be restricted to high levels above the pool surfaces and the authors concluded that: 'It is not correct, therefore, to regard such species as *Calluna, Eriophorum vaginatum* and *Trichophorum* as plants of the drier phases of bog growth, but as with the Sphagna, the habit and growth-form of these species is a good reflection of the degree of humidity under which they exist' (page 426).

Another point of note was the very high percentage of 'flat' communities (0-15 cm above water table) and the moderate percentages of pools in all six bogs that make up the Silver Flowe, as well as the extremely low percentage of tall hummocks. The bogs, which are morphologically of 'blanket bog' type but with distinct 'raised bog' affinities, seemed to be in an actively growing state and 'recovering' from a recent rise in the water-table as was shown in the small pits which Ratcliffe and Walker dug in a transect across one of the bogs, Snibe Bog. These pits, 67 in all, were dug in pairs to cover adjacent hummocks and hollows or a single pit was dug on the 'flats', at the quadrat points used to describe the vegetation. Their purpose was to 'elucidate the status of the pool and hummock complex' the authors having noted previously, from peat borings, that 'the most striking feature of the stratigraphy of both bogs (Snibe Bog and another of the series, Craigeazle Bog) is the absence of any trace of lenticular structure suggesting 'regeneration complex' in the past, except in the most superficial layers'. The pits clearly showed a surface of fairly well humified *Sphagnum-Molinia* peat underlying the present surface of the Snibe Bog at a depth of less than 40 cm in most cases. (As will be seen later, this is very similar to the stratigraphy of large parts of Bolton Fell Moss.) The 'regeneration complex' was therefore a very recent feature and only in a few cases (three quadrats out of the 27 on the bog plane) were there indications of a pool having formed over a hummock.

Ratcliffe & Walker's paper is an important landmark. It provides an excellent record of the vegetation of a series of fairly intact ombrotrophic bogs, and for the first time, some 35 years after Osvald's monograph, there was a determined attempt to look at the stratigraphy underlying a 'regeneration complex' situation. It also led on to further work by one of the authors (Walker & Walker 1961) in an attempt to resolve the question of lenticular regeneration and detailed further studies on the same bogs (e.g. Goode 1970, Boatman & Tomlinson 1973).

2.5 EXPANSION AND SPECIALISATION 1960-1977

Since 1960 there has been both a notable expansion of activity in bog ecology and a tendency for research to be specialised on a particular aspect of the field. At the same time a number of new techniques were introduced – notably radiocarbon dating – and old techniques became more widely and precisely used, such as macrofossil and rhizopod analyses. Specialisation showed itself throughout the period with papers concerned with aspects such as the chemistry or hydrology of bog systems rather than comprehensive description, and, increasingly, with papers centred on the ecology of a single bog species or group of species. Many of these 'specialism' papers are germane to the topic of this thesis but are more appropriately dealt with in the Discussion (section 6) – for example, Clymo's studies on the growth of Sphagna (1963, 1965, 1970, 1973), Ingram's review of mire hydrology (1967), Wein's study of *Eriophorum vaginatum* (1973) and work such as that of Boatman & Tomlinson (1973) on the detailed structural and hydrological features of bog surfaces.

The first attempt at a stratigraphic approach to the problem was that of the Walkers who examined the sections left by peat cutters in a number of Irish bogs in the summer of 1959; two papers emanated from this work. The first (Walker 1961) was given as a lecture in a Symposium on Quaternary Ecology held by the Linnean Society in December 1959 and draws broad conclusions from the observations of the previous summer's work. Details of the sites, the methods used, the stratigraphic sections and rather more firm conclusions are given in the second paper (Walker & Walker 1961).

All the sections studied were at some distance from the original bog margin and therefore probably related to the former 'bog plane' rather than sloping 'rand' communities. They distinguished five distinct types of peat which were recorded using a wire grid of 10 x 10 cm squares and they showed finer, more exact detail than any previous work. These stratigraphic observations were backed up by two series of macrofossil analyses of peat monoliths taken through particularly interesting parts of the sections, the stratigraphic stages being loosely linked by pollen analyses. Although 'a rather crude system designed only to obtain an unsophisticated answer' (D. Walker, personal communication) their observations led to conclusions that shed much doubt on the validity of the cyclic theory.

They noted (Walker & Walker 1961, pp.182-4) that 'hummocks originated primarily on *mature* surfaces' and that these hummocks 'usually persisted when the rest of the surface was overgrown by *Sphagnum* and may have limited the extent of pools . . .' These hummock survivals may then be rejuvenated and survive further flooding of surrounding areas, meantime acting as centres for the dispersal of hummock plants over the surrounding lawns of Sphagna; their eventual fate was either to be incorporated in a later 'mature surface' or to be overgrown by '*Sphagnum*-rich communities during a phase of very active peat accumulation'. They saw very little evidence of autonomous degeneration of hummock tops and concluded that 'the most interesting feature of most of these sections . . . is the evidence for a bog surface reacting all over and in the same direction to a change in conditions'. This naturally led them on to say that 'there were few, if any, certain signs of the operation of an internally controlled, automatic, cycle in which pool formation and hummock degeneration were discrete stages', and that the main system by which the bogs have regenerated was *via* periodic rejuvenation due to an increase in wetness of the bog surface . . . 'whilst the maturation sere, progressing towards stability, developed under constant or drying conditions'. This sort of process, they recognised, is akin to that resulting in recurrence surfaces 'although their stratigraphic manifestation is not so distinct'.

The Walkers also recognised what they termed a 'short-cycle' regeneration complex in peat which was without distinct pools and approaching maturity. In the symposium paper (Walker 1961, p.31) this is described as being 'Occasionally . . . found in contexts where it is reasonably certain that neighbouring hummocks and hollows have been converted, the one to the other, each providing a source of new species for the other'. In the joint paper the emphasis is changed somewhat – '. . . there was considerable evidence for a 'short-cycle' regeneration complex . . . a fluctuation in the relative importance of *Sphagnum, Calluna* and *Eriophorum vaginatum* from time to time in a given place . . .' (Walker & Walker 1961, p.184). This is ascribed to some sort of periodicity in the life history of the dominant plants rather than to a self-regulating regeneration cycle.

While this work is undoubtedly important as a pioneering study of microstratigraphy it is lacking in some important respects and was not followed up due to the author's emigration to Australia. Perhaps the most vital lack is that of independent evidence of climatic change and the associated absence of any real chronological control of the sections. Stages of formation of, for example, *Sphagnum cuspidatum* peat at different points in the same section could not therefore be linked in time or related to a known change to a cooler and/or wetter climate. Similarly there is a general lack of information on the species involved, only three small monoliths being analysed for macrofossils, and at that time (1959-60) much less was known of the ecology of individual bog species and of the processes involved in peat formation such as the importance of the 'sulphide layer' (Clymo 1965). One may also note that because of the derelict nature of the bogs investigated the stratigraphy could not be related to a modern surface as is possible in a recently cut bog such as Bolton Fell Moss. Nevertheless, the Walkers' observations serve as a kind of starting point for all recent research in regeneration and are referred to in virtually all papers published since 1961.

In his regional study of bogs in southern Finland Eurola (1962) accepts the validity of cyclic regeneration, citing both Tansley (1939) and Kulczynski (1948) and reproducing the latter's diagram. This study is the last one to accept the theory in this rather uncritical way; thereafter the Walkers' observations are seen to institute a more critical awareness of the problems involved in peat stratigraphy.

Overbeck (1963) was the first investigator to bring these problems to the fore in a very readable introduction to a lecture course on the 'Science of Moorlands' published with other short, informative papers (e.g. Schneekloth 1963, Grosse-Grauckmann 1963) on the same theme.

While not denying the possibility of a self-regulating cycle he points to various pieces of evidence which have made him disinclined to accept it. These include the 'false idea that the hummock is drier than the hollow' whereas the opposite is often true during summer drought due to the anatomical structure of the different *Sphagnum* species, the close-packing of the *Sphagnum* stems in hummock situations, and the shade afforded by the *Calluna* and *Eriophorum vaginatum* of the hummocks (Overbeck & Happach 1957). Overbeck also considers the question of scale and form with regard to the recognition of former surface features, such as low hummocks (30 cm high), in the stratigraphy and points to the lack of any stratigraphic proof of cyclic development in the profiles of north German bogs. Following Walker (1961) he notes the presence of 'lightly waved' and flat *Sphagnum* carpets which seem to have persisted for some time in north German bogs, together with long-lived hummock structures.

As previously mentioned the early 1960's saw a considerable expansion of research into bog development and the ecology of bog species. Radiocarbon datings of stratigraphic changes such as recurrence-surfaces were increasing in number and papers such as Godwin (1960), Lundquist (1962) and Nilsson (1964) re-opened the debate on their age, number and synchroneity. Of particular interest to the regeneration complex question was the clear recognition and dating of an uppermost recurrence surface (or surfaces) in late-Medieval times and later – though well within the period for which independent climatic evidence becomes available, this connection was not explored to any great extent. Papers by Chapman (1964a), Tallis (1964b) and Clymo (1965), among others, also provide much useful information. Chapman, working on a site only 14 miles from Bolton Fell Moss, describes two types of vegetation distinguished by their proportions of *Sphagnum magellanicum* and *S. papillosum. S. imbricatum* also occurs widely but there is no distinctive hummock-hollow complex. These results are of interest in considering the macrofossil analyses from Bolton Fell Moss in which the same species were found. Tallis, in the third of his interesting 'Studies on southern Pennine peats', notes that 'the typical lenticular arrangement of a regeneration complex has, however, never been observed', in his area, and has various observations on recurrence surfaces and the behaviour of *Sphagnum imbricatum.* Clymo's experimental study of the breakdown of *Sphagnum* in peat has obvious bearing on this study and all others on peat stratigraphy since, for example, he found the rate of breakdown of *Sphagnum papillosum* to be only half that of *S. cuspidatum.* The results of this and of other related work since 1960, are of value in the interpretation of results from the present study and will be returned to in the discussion (section 6).

Also in 1964 the first textbook source of doubt regarding the cyclic hypothesis appeared – Ratcliffe's chapter in Burnett's *'Vegetation of Scotland'.* Ratcliffe notes that the '. . . idea is perhaps too facile, and has not so far been borne out by careful stratigraphical examination . . .', and that though conversion of a hollow into a flat or hummock may occur in some bogs '. . . they are persistent features which have occupied exactly the same places for a long time'. Naturally he is drawing mainly on his knowledge of the Silver Flowe bogs (Ratcliffe & Walker 1958) but also on observations from elsewhere, such as the bog near Kinlochewe (Gimingham *et al* 1961). Other textbooks written since 1960 either ignore the topic, probably because of its complexity and the controversy surrounding it, or else, as in Kershaw's *Quantitative and Dynamic Ecology* (1964) the cycle is given prominence as an 'outstanding example' of the hummock and hollow cycle (linked to others such as Watt's grassland cycle) and proved by the stratigraphic work(!) of Oswald (1923) (sic) and Godwin & Conway (1939). The second edition (Kershaw 1973) repeats this view *verbatim,* despite the *caveat* of Proctor, in Tansley 1968, already noted.

Faegri & Iversen's formative *Textbook of Pollen Analysis* (1964) takes a rather confusing line on the regeneration issue. They note (p.44) that '. . . among the terrestic peats one series deserved special comment, viz. the regenerative types, which are in Northern Europe represented by some *Sphagnum* peats'. They outline the cyclic succession according to von Post & Sernander (1910), referring to the stratigraphy produced as 'bacon peat' – a rather apt description, as figure 2 shows – but then go on to say that Walker & Walker (1961) '. . . have published detailed sections of Irish peat walls with regeneration structures. On the whole, a detailed analysis confirms earlier opinions as to the genesis of such peats, though both hollows and hummocks appear to be more permanent features than hitherto assumed'. This does not really seem to accord with the conclusions of the Walkers. Faegri & Iversen also point out the role of *Sphagnum fuscum* in the 'classical regeneration cycle' and speculate that some of the differences observed in Ireland may be due to its absence. However, they do not mention Osvald's 1923 work, and in the Komosse bogs *S. fuscum* is quite rare, Osvald classifying it as a species of the 'eastern facies' of raised bog. Faegri & Iversen then go on to accept the theory . . . 'By the cyclic regeneration the bog grows up . . .' and the same view is put forward in their third edition (1975) with only the addition of a mention of Casparie (1969) '. . . who queries the existence of a cyclic regeneration system in Dutch bogs'. While accepting that the primary purpose of the book lies elsewhere it is disappointing not to find a fuller treatment of peat formation in such an important palaeoecological textbook.

Hansen (1966), working on a remnant of the central plane of Draved Kongsmose, southern Denmark (3.5 hectares remaining from an original 500 hectares) comes to conclusions similar to those of Walker & Walker (1961). From investigations of surface vegetation, a number of short cores and a 200 x 60 cm section constructed from nine further cores, he concludes that the hummocks '. . . seem to be perfectly capable of producing peat in pace with the hollow communities . . .' and that the successions represented in figure 8 should be regarded not so much as cyclical but as the directions in which the vegetation would change in response to a greater or lesser degree of wetness on the bog surface – that is, such diagrams (including Osvald's) should be viewed as *zonation* diagrams. The paper also includes information on *Sphagnum* growth rates related to recent climate (1953-1956) and some chemical analyses. On figure 8 the solid lines are regarded as almost certain successions and the important roles of *S. magellanicum, Calluna* and *Eriophorum vaginatum* are clearly shown. Broken lines show only 'possible tendencies' which were not clearly demonstrable in the peat sections. The relevance of this succession diagram to the situation at Bolton Fell Moss will be obvious from the results in section 5.4.

Tolonen has produced a number of papers on bog development and vegetational history in Finland. Two of his papers deserve comment, published in 1966 and 1971. The former is notable for the use made of fossil rhizopods to indicate past hydrological conditions which, combined with macrofossil analyses, give a rather fuller picture of bog development. This 1966 paper describes the Flandrian development of the bog as a whole and has little on the regeneration complex *per se,* though Tolonen feels able to conclude (p.164): 'Stratigraphical indications (three widely-spaced boreholes) prove that a

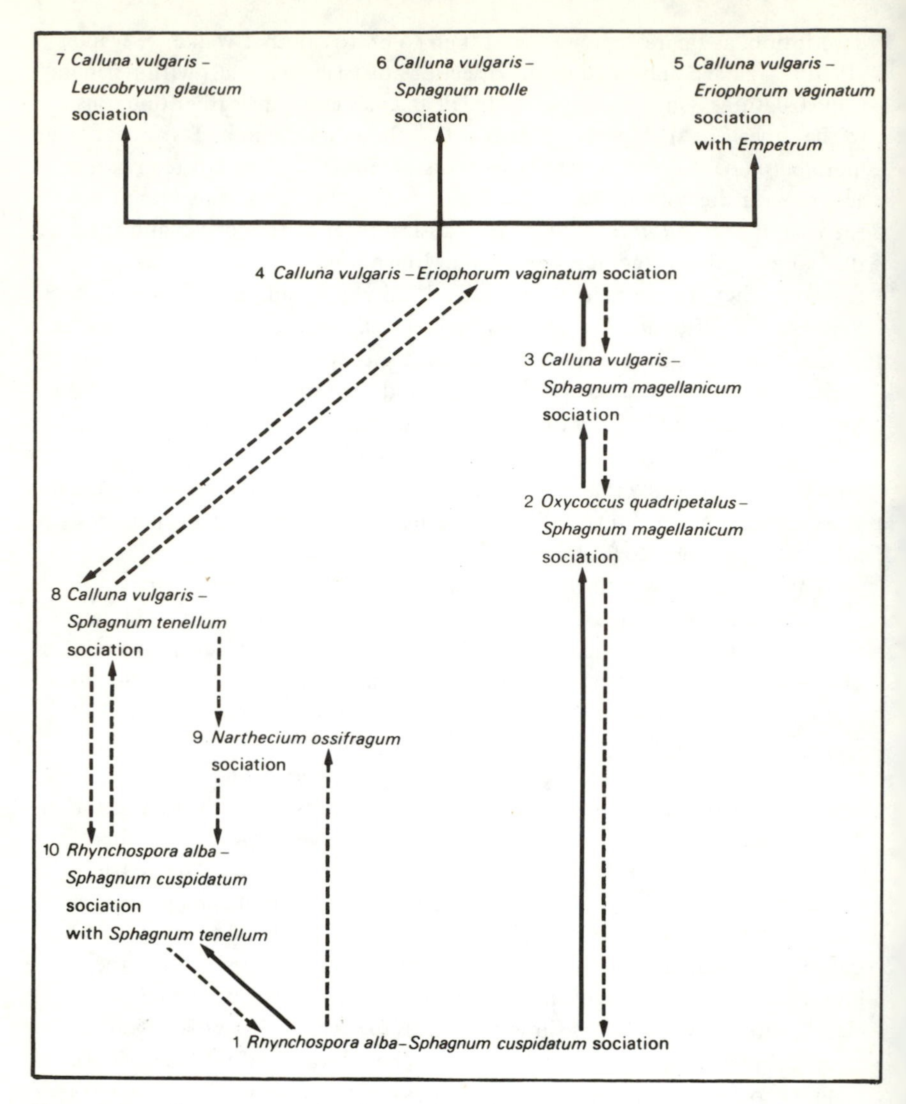

Figure 8. Zonation and succession diagram from Draved Kongsmose (after Hansen 1966).

regeneration cycle with alternating hollows, even water pools, and hummocks prevailed in Varrasuo later in the Atlantic period'. This statement seems to be at odds with Tolonen's acceptance (p.156) of Walker & Walker's (1961) strictures on the use of single boreholes rather than exposed peat faces. In his 1971 paper, Tolonen notes that his earlier stratigraphic analyses (the 1966 paper mentioned above, his 1967 work on north Karelian bog development and his 1968 paper on the development of the central Finnish bogs) '. . . suggest cyclic alternation of smaller hummocks and hollows . . .'

Tolonen (1971) is specifically on 'bog regeneration' which he takes to be 'cyclic alternation of hummocks and hollows' (p.143). At Klaukkalan Isosuo in S. Finland it was possible to record stratigraphy from open peat sections and five vertical profiles, from three faces, were analysed using a variety of methods. These included pollen, macrofossil and rhizopod analyses and also calorimetric analyses performed using a bomb-calorimeter. The bog is somewhat peculiar in consisting almost exclusively of little humified *Sphagnum fuscum* peat (with small amounts of *S. magellanicum, S. papillosum* and a few other species) except for very thin highly humified 'streaks' of jelly-like, slippery material thought to be 'mucous lichen residues', often containing *Calluna* fragments and sometimes rich in *Eriophorum vaginatum* remains. Most of these streaks are only a few millimetres to a centimetre or so thick and up to 30 of them were counted during borings through the bog. Tolonen includes three photographs and two stratigraphical diagrams of the uppermost 80 cm of the peat faces containing these streaks, some of which can be traced for several tens of metres in an anastomosing fashion. He notes (p.152) that '. . . they undulate, obviously following the ancient microtropography of the bog surface' and in some places they terminate on meeting a hummock, rising upward to meet it. There are few hummocks in the stratigraphy though, and in general the stratigraphy is quite unlike that of most raised bogs, at least in western Europe, though Tolonen claims to have seen similar streak systems in a number of bogs in Sweden and Germany as well as other bogs in Finland. Their origin is not at all clear though Tolonen believes that there are similiarities with the 'short-cycle regeneration' of Walker & Walker (1961) in that the hollow phase of the cycle is absent – peat formation comes to a 'complete standstill' during streak formation and then begins again with the deposition of lightly humified *Sphagnum fuscum.* Tolonen speculates (p.164) that they may be caused by minor climatic fluctuations or by the lifecycles of the dominant plant species (following Walker & Walker 1961), but as he has no precise datings or independent climatic evidence these possibilities remain to be clarified. The extreme thinness of the streaks (a few millimetres up to a centimetre) lead one to wonder whether one or two extremely dry summers, combined perhaps with 'flash' fires, might not offer another explanation. The periodicity is thought to be about once in 90-150 years so that such an explanation is climatically possible, though no fine charcoal seems to have been observed by Tolonen. Weight is added to this possibility, and the possibility that man may have played a part in the formation of the streaks, by Tolonen's observation that they almost all occur in the Sub-Atlantic peat (500 BC onwards) and that the rhizopods associated with the streaks are all of the 'tyrfoxene' group (Grospeitsch 1953) '. . . which is 'not indigenous to bogs' but, instead, characteristic of drained peatlands' (p.155). Obviously any overall theory of bog

17-9 100 BC

17-8 300 BC

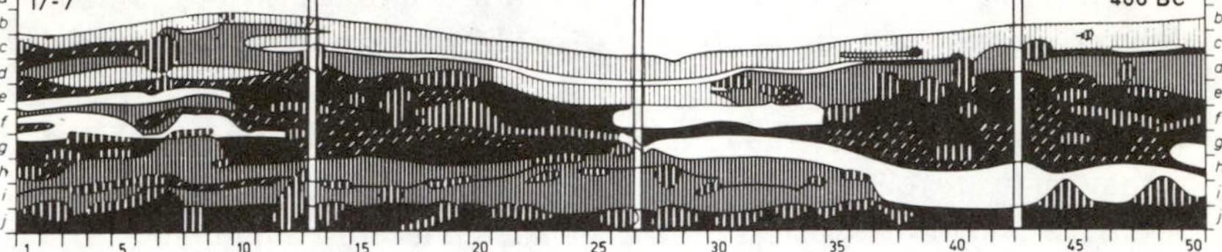

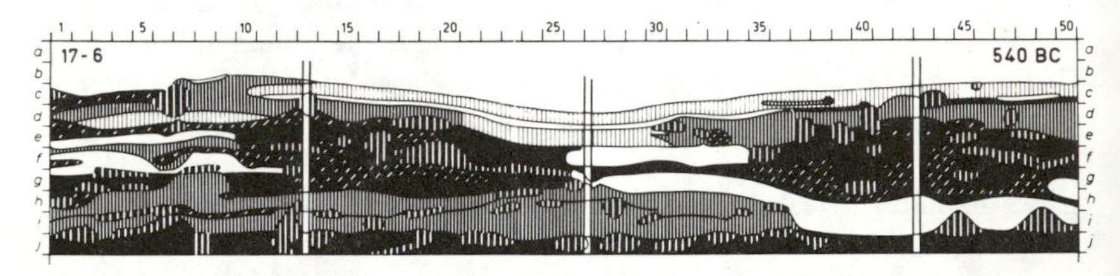

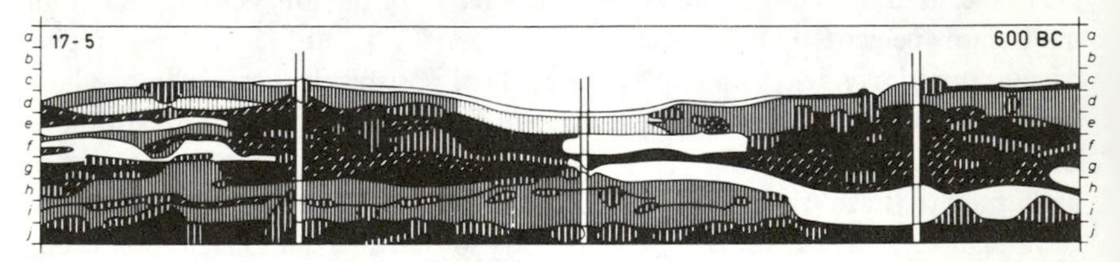

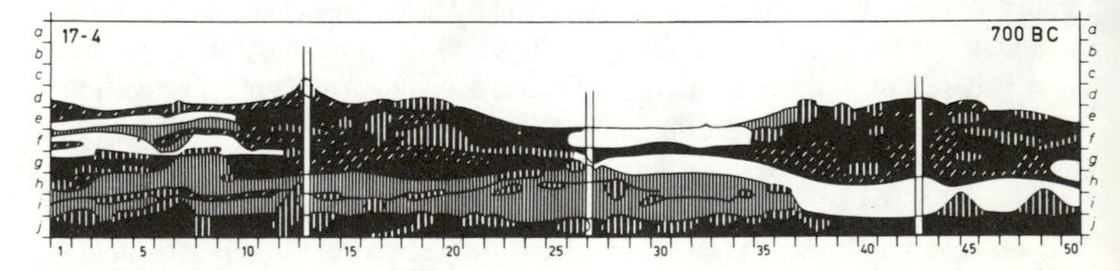

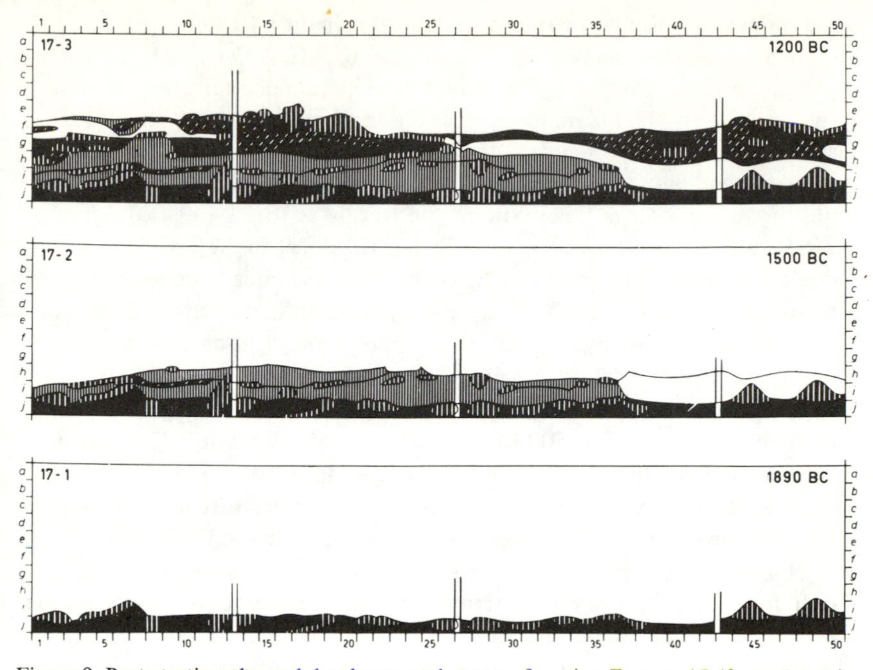

Figure 9. Peat stratigraphy and developmental stages of section Emmen-17 (from Casparie 1969).

growth must attempt to take into account observations such as those of Tolonen (1971) but in the absence of any evidence for widespread hummocks and hollows and in the overall dominance of a single *Sphagnum* species, Klaukkalan Isosuo presents a rather unique case and may not fit into a general model.

Two of the most significant recent papers on bog development are those of W.A. Casparie (1969, 1972). He spent more than a decade, from 1959, investigating part of the vast peatland area of the Netherlands/German border region known collectively as Bourtanger Moor, some 80 km long by 15-20 km wide, focussing his attention on the area to the east of Emmen. The 1972 publication comprises his doctoral thesis *in toto* and is concerned with everything from the Late-Glacial deposits to recent erosion phenomena. However, it is not necessary to refer to this large work, which is in English, as the shorter German publication of 1969 deals specifically with the hummock-hollow complex in the area where it is best developed in the stratigraphy, section Emmen-17. This section, 5 m long by 1 m deep, was recorded by Casparie using the methods of Walker & Walker (1961) and three monoliths taken for pollen, macrofossil and rhizopod analysis. Using the pollen analyses to establish ten broadly synchronous zones, Casparie presents the development of the peat section in nine stages, from 1930 BC to 100 AD (figure 9).

In a detailed discussion of the succession in section 17 (pp.173-178), Casparie points out that the rejuvenation of the mature bog surface post-1930 BC (phases 17-1 to 17-2) is similar to that observed by Walker & Walker (1961), and is due to climatic change, affecting, as it does, large areas of the Emmen bog-complex. Subsequent changes, however, he ascribes not to climatic causes but to local hydrological conditions, specifically to the waxing and waning of a 'raised-bog lake' (*Hochmoorsee*) which formed just to the east of section 17 due to the convergent growth of a number of domed bog complexes. Following rejuvenation then (phase 17-2) we have a period of 'maturation' and succession to drier communities by the infill of the pool at the right-hand side of the section (phases 17-3 and 17-4), then an increase in wetness due to the rising level of the lake (17-5 and 17-6) giving less humified peat. This bog lake overflowed catastrophically at about 500 BC, though the area of section 17 remained wet and accumulated slightly humified peat except in the centre where highly-humified *Sphagnum-Calluna* peat accumulated – apparently in a hollow though Casparie notes the possibility of subsidence due to the bog-burst.

Not surprisingly from the above development Casparie states '. . . we think that there are no simple connections between the amount of rainfall and the creation of changes in humification'. He goes on to point out how many inter-related factors there are in peat formation, and goes on to say that from his evidence, '. . . there can be no question of autonomous cyclic processes. The very long life-span of the hummocks themselves, often more than 1 000 years, cannot possibly be fitted into a cyclic system'. (This last point had by that time also become clear from the studies on Bolton Fell Moss.)

While Casparie's work is clearly of importance, especially in the way it backs up several of Walker's conclusions, it also suffers from an inability to link stratigraphic changes with independently-known climatic changes. This inability stems from the fact that during the 19th and 20th centuries the bog east of Emmen was repeatedly burnt and cultivated for buckwheat (*Fagopyrum esculentum,* a cereal substitute) resulting in the loss of the topmost metre or so of peat, dating from about 100 BC onwards. Taking this into account, and the rather special case presented by the influence of the raised-bog lake, it is not surprising that Casparie felt unable to ascribe changes in peat humification and type to climatic change, apart from the 'alteration of the original balance' at around 2000 BC.

The theme of bog hydrology is undoubtedly of importance in providing some explanation of species pattern, growth rates and perhaps the fate of individual hummocks and hollows. There is a good deal of hydrological information in a variety of such thematic studies published since 1960. These include Romanov's *Hydrophysics of Bogs* (1966); Ingram's review of 'Problems of hydrology and plant distribution in mires' (1967); Boatman & Tomlinson's study of bog structure and hydrology (1973); Goode's unpublished thesis on, particularly, the water relationships of some Silver Flowe bogs (1970) and Clymo's 1973 study of *Sphagnum* growth. Information from such studies, together with that brought together in Moore & Bellamy's book *Peatlands* (1974), is more appropriately introduced at the stage of discussion of the results from Bolton Fell Moss and is therefore left until section 6.

Finally, amongst the great number of papers produced since 1960, one may note three recent additions which approach the topic of bog growth from a historical viewpoint – Aaby & Tauber (1974), Dickinson (1975) and Aaby (1976). Dickinson's paper, on recurrence surfaces in Rusland Moss, contains useful information on the two latest recurrence surfaces (RYI:1200 AD; RYII:400 AD) but without much stratigraphic detail (the bog is not being cut), and almost no reference to climatic change. Though cyclic alternation of hummocks and pools is claimed to have occurred in one small area the evidence is not conclusive – no stratigraphic detail could be obtained and the claim is based on only a few macroscopic remains and on growth rates deduced from small changes in pollen content. These deficiencies are, to be fair, recognised in the statement (p.926):

> Though it has not been possible to relate the phases more closely, it seems that the smoothly oscillating pattern of peat growth is linked with the pool and hummock succession.

One further point which deserves mention is the suggestion that the regrowth of peat above a 'standstill' level could be due to impeded drainage in humified peat, a factor which has been noted by others such as Tallis (1973), Ingram (1967) and Godwin (1954).

Aaby's work on peat formation and climatic change is of great interest. He was fortunate enough to have done 70 radiocarbon dates, of which no less than 55 were from a single profile through 2,5 m of peat from Draved Mose. From these dates and analyses of humification and some macrofossil and pollen data, Aaby (1976) claims to have found cyclic climatic variations with a periodicity of about 260 years over the last 5,500 years. These variations seem to be statistically valid. Not much is said about the regeneration complex theory except to point out that, as some hummocks in Draved Mose have a life span of over 2,500 years, cyclic alternation appears to be unlikely. In the joint paper (Aaby & Tauber 1974) the long-contested relationship between humification and climate is resolved – the 'orthodox' view of Granlund (1932) and many others, that peat of low humification is formed in a wetter climate than peat of high humification, is upheld, against the views of, for example, Olausson (1957) and Schneekloth (1965). Casparie (1969, 1972), is not cited by Aaby – perhaps a recognition of the peculiar circumstances of peat formation in the Emmen bog. Autocompaction of peat is also discussed and much information is given on accumulation rates in different local conditions. The details of such work are more appropriately discussed with relevance to the data from Bolton Fell Moss which follow in the next three sections.

2.6 SUMMARY OF PREVIOUS RESEARCH

After a long period of speculation and imprecise observations on the peat-forming process and the growth of bogs, a superficially attractive theory involving cyclic alternation of hummocks and hollows building upwards to form domed bogs was postulated in the early 20th century. Apparently confirmed by studies of surface pattern and the ecology of bog species this theory became the widely disseminated orthodoxy of the first half of the century and was not formally tested until 1961. The doubts exposed at that time have been reinforced since by a number of studies of bog growth though there are still a number of adherents to a modified form of the theory. Related studies of hydrology and the ecology of individual species have added greatly to our general knowledge of bog ecology. The importance of climatic change has been recognized and the problems of diagenesis in peat formation clarified. The present study, building on a critical evaluation of previous research, aims to correlate changes in peat stratigraphy and species composition on a micro-scale with known climatic changes of the last two millenia and so test the central postulate of the regeneration complex theory, that of an autogenic alternation of hummocks and hollows.*

*Since this text was written three recent German-language textbooks, which summarize aspects of peat stratigraphic research have come to author's attention:-Overbeck (1975) *Botanisch-geologische Moorkunde;* Ellenberg (1978) *Vegetation Mitteleuropas mit den Alpen in ökologischer Sicht;* and Göttlich (1980) *Moor- und Torfkunde.* They do not materially alter the conclusions reached here.

3. SITES INVESTIGATED

3.1 BOLTON FELL MOSS

3.1.1 Suitability

As will be apparent from section 2 one of the main pre-requisites in testing the regeneration cycle theory was felt to be the ability to relate peat stratigraphy to independently assessed climatic changes. This can only be done in a bog which has grown without interference up to the present day but which, paradoxically, is in the process of being destroyed by extensive peat-cutting. This enables one to observe lateral variations in the uppermost peat which may pass unnoticed or cause correlation difficulties in a borehole study. A further condition is that the peat must be cut by hand rather than machine so that the finer details of micro-stratigraphy may be recognized. Machine-cutting causes compression and distortion of the soft recent peat and smears the cut section with peat from different levels to a much greater degree than hand-cutting. Bolton Fell Moss met these conditions, at least until 1970. Hand-cutting was practised from the start of operations by the Boothby Peat Company in 1956 until the introduction of small machine-cutters in 1968, followed by large-scale mechanisation in the early 1970's.

The depth of peat cuttings and a rapid rate of peat accumulation in recent centuries are two further features of importance. Slow peat accumulation will obviously limit the scale of climatic and pollen-analytical event that may be 'resolved' in the stratigraphy, while shallow cuttings will limit the time period available for study. In both these regards Bolton Fell Moss is more suitable than any other site examined and the extensive nature of the peat company's workings meant that many miles of section could be examined.

3.1.2 Location and surface features

Bolton Fell Moss covers almost 4 km^2 of Ordnance Survey One-Inch Sheet 76, centred on grid reference NY4969. The location map (figure 10) shows the site in relation to some of the more important other mosses in the Solway Lowland,

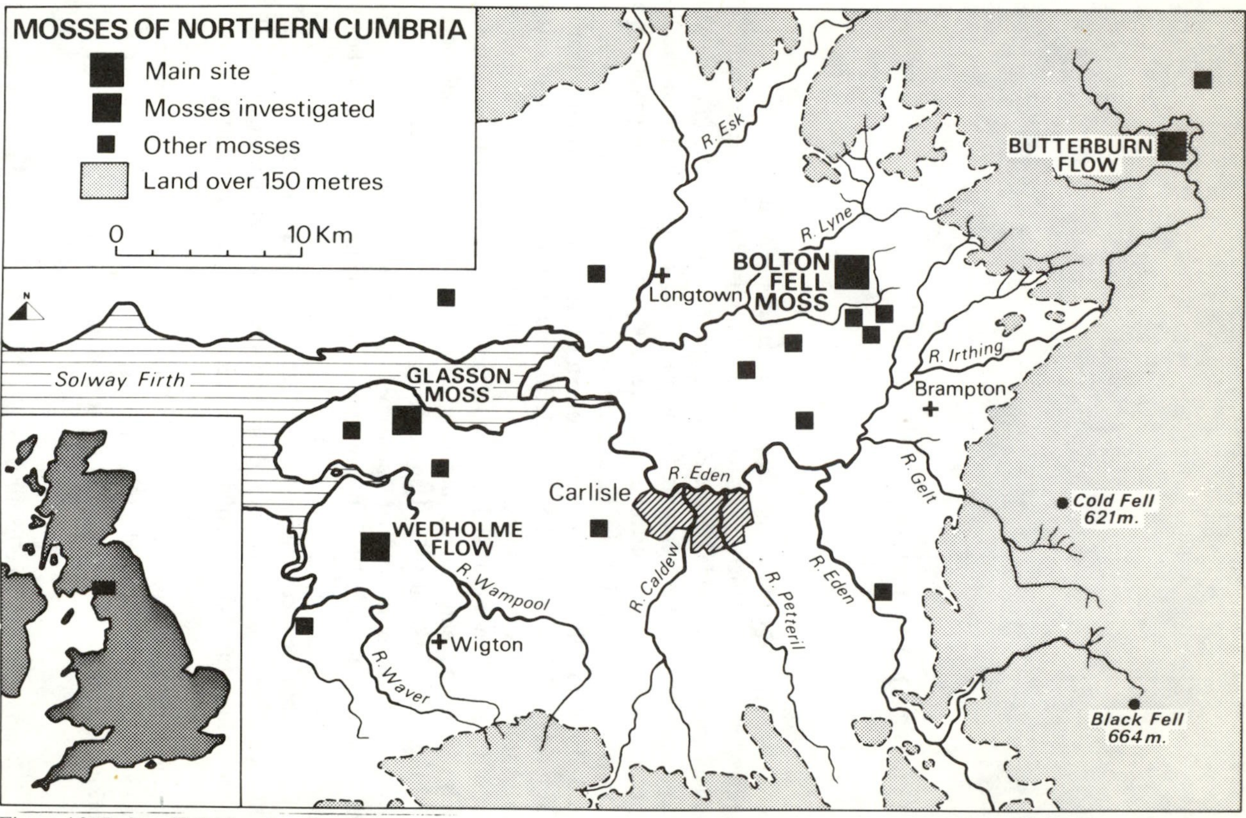

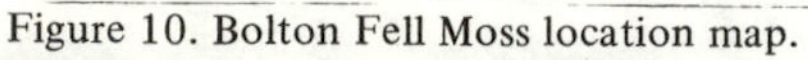

Figure 10. Bolton Fell Moss location map.

all of which have developed over the deposits of the last glaciation (Devensian Stage) – the underlying solid geology of Carboniferous and Permo-Triassic rocks has had little effect. This Bewcastle drift sequence of till, glacial sand and gravel and glacial lake alluvium is now all thought to date from the later stages of the Devensian (Day *et al* 1970) and the reality of the Scottish Readvance glaciation (Trotter & Hollingworth 1932, Walker 1966) is now in doubt. The drift covers most of the area below 750 ft (229 m) extending locally to over 1000 ft (305 m) and its irregular surface gave rise to many ill-drained basins and some kettle-hole lakes. Hydroseral development throughout the post-Glacial period (Flandrian Stage), has culminated in a number of raised-bogs, locally known as 'mosses' or 'flows', contrary to the postulated development of oakwood (Tansley 1939), as Walker (1970) has demonstrated.

A number of these Cumbrian mosses have attracted research workers including Godwin, Walker & Willis (1957), who determined the radiocarbon ages of the pollen zones established earlier by Godwin (1940), by dating a

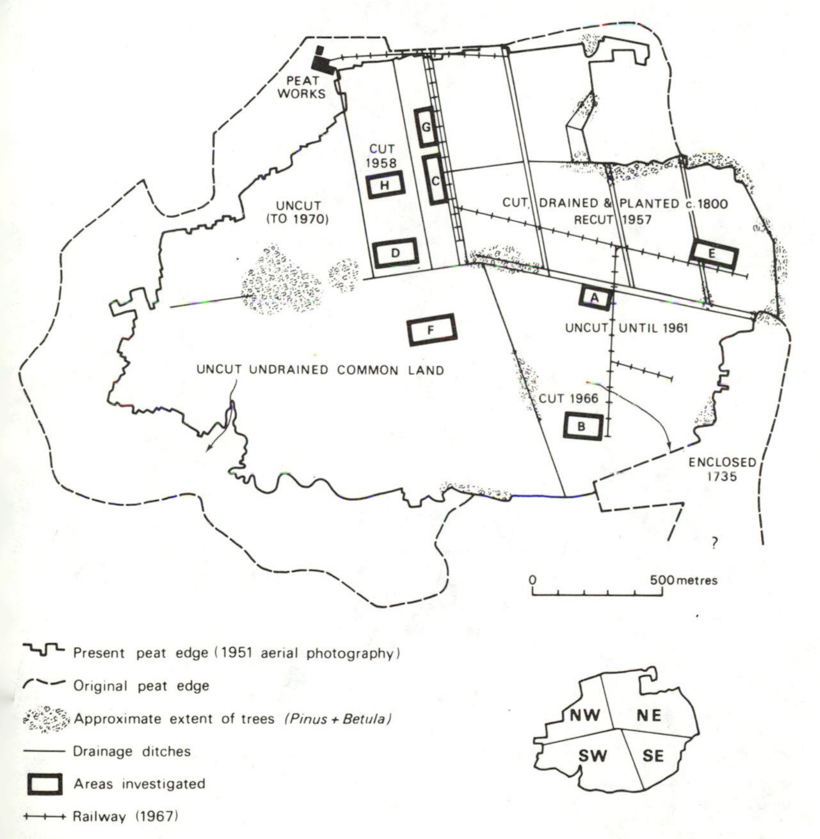

Figure 11. Bolton Fell Moss surface features and peat areas studied in detail.

pollen diagram from Scaleby Moss, some 8 km south-west of Bolton Fell. Chapman's work on Coom Rigg Moss (1964a, b, c) has already been mentioned and Walker (1966) worked on the palaeoecology of a large part of the Cumberland Lowland to produce a more or less definitive vegetational history of the area up to Roman times, but with little or no detail thereafter in what were full Flandrian pollen diagrams constructed according to the older tradition which did not sample the uppermost layers at all, or at least not closely.

The surface features of Bolton Fell Moss are shown on figure 11, and the aerial photographs of 1951 and 1974 (plates II and III) and the author's photographs (plates IV and V). The moss is sub-circular in outline though the original edge of the peat covered area has been lost by centuries of piecemeal cutting and, in the north-east quadrant of the moss, by a probable 'bog-burst' (1951 aerial photography). Such a bog-burst has been recorded in some detail from Solway Moss (Walker 1772). The presumed original margin of the bog is shown on figure 11, reconstructed from indications on the aerial photographs,

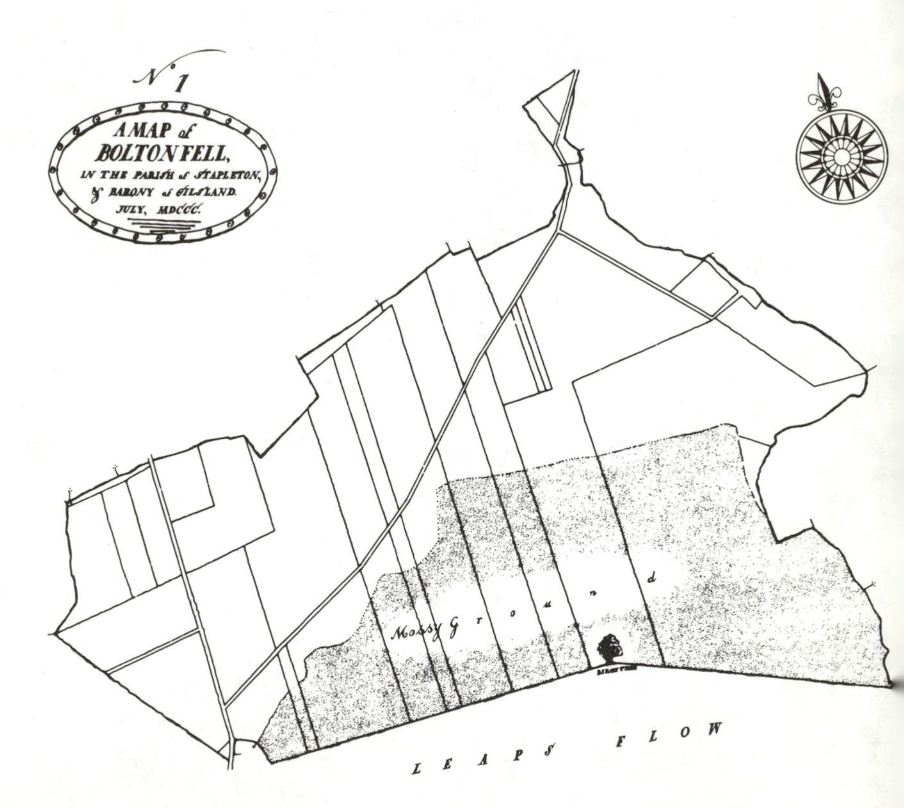

Figure 12. Bolton Fell Moss: map of the reclamation attempt, 1800 A.D.

from a consideration of the contours of the surrounding land and from field observations. Systematic peat-cutting and drainage and planting works associated with an attempt to reclaim the northern half of the moss in 1800 AD are known from a parchment map lodged in the County Record Office, Carlisle, partially redrawn and presented as figure 12. The effects of this are clearly seen on the 1951 aerial photographs (pre-dating the establishment of modern peat-working) and the central clump of trees also almost certainly dates from this 1800 AD disturbance. Slight alteration of the westernmost ditch of the two ditches running north-south to the central clump of trees by the Boothby Peat Company caused a major setback to this research in 1969. This affected what was then the main area of investigation (Area C, figure 11), but it will be more relevant to deal with the history of peat-cutting when discussing the results from Area C.

The gross topography of Bolton Fell Moss has now been much altered from its original state. The surface of the north-east quadrant has been considerably lowered by the 19th century and present-day peat workings. Shrinkage of drained peat is a well-attested phenomenon (Holme Post in the East Anglian Fens shows a shrinkage of 3.5 m over a century) and even the modern workings have produced distinct 'steps' in the surface level of the bog as is evident on the photographs (plates IV and V). Although the highest surface today is found in the undrained south-west quadrant this may therefore be an artifact. For this reason, and because the time involved would not, it was thought, be repaid by useful results, no systematic boring and levelling was undertaken. There is no doubt that the moss as a whole is ombrotrophic and, as will be shown from the macroscopic analyses, it has been so for at least the last two millenia. Small areas do show a minerotrophic influence either due to the release of nutrients by drainage or due to proximity to the central copse which is on a boulder-clay inlier. The tonal gradations on both aerial photographs are evidence of water movement and relative saturation, amongst other things, which must be related to the underlying mineral ground as well as to drainage ditches.

While, as already noted, no systematic boring programme was attempted, the depth of peat was ascertained at numerous points and in all areas where peat sections were recorded. It was hoped that these might distinguish any separate basins of accumulation and underlying ridges such as found by Chapman (1964a) at Coom Rigg Moss and Pigott & Pigott (1959) at Malham Tarn Moss. Sample peat depths include the following:

Area A:	7.50-7.85 m	Area E:	4.20 m
Area B:	5.10 m	Area F:	10.60 m
Area C:	8.60 m	Area G:	7.30 m
Area D:	7.90 m	Area H:	6.00-7.10 m

The general height of the peat face exposed around the cut margins of the bog averages 1-2 m. These depths are at variance with that quoted by Walker (1966, p.16):

Another type of bog occurs on convex slopes and the flatter hill tops even at comparatively low elevations (e.g. Bolton Fell, 110 m (361 ft) OD) and must owe its maintenance to the intense oceanicity of the climate. The peat is only about 3 m thick at the deepest points and cutting and erosion around the edges have effectively modified the surface vegetation so that *Calluna vulgaris* and *Eriophorum vaginatum* now share the dominance while *Sphagnum* spp. play a comparatively small role.

This peat depth and description of the vegetation of Bolton Fell Moss can only refer to a small area to the north and south of the central copse, within the north-east quadrant, which does superficially appear to be the crown of the bog. Elsewhere Sphagna are abundant and the shallowest depth recorded by the author, apart from immediately around the central copse, is the 4.2 m noted above from Area E – where the peat has been cut and undergone shrinkage. (As this quotation from Walker is the only literature reference to Bolton Fell Moss, it is felt important to correct it as a matter of record.)

Separate basins within the bog have not therefore been discovered by the 20-odd full borings performed by the author and although of course they may still exist there is no doubt of the ombrotrophic nature of the peat-forming communities during the time-period on which this study is based.

3.1.3 Present vegetation

The vegetation of the uncut south-west quadrant of Bolton Fell Moss is an especially fine example of a Sphagnetum dominated by *Sphagnum magellanicum.* Even more interesting is the probability that this species has only recently achieved dominance over a former cover of *S. imbricatum* and *S. papillosum,* as will be discussed in later sections (5 and 6). At first sight the vegetation in this quadrant appears monotonously uniform – a more or less flat surface dominated by *Eriophorum angustifolium* (plate IV) with occasional 'islands' of *Calluna vulgaris* and *Eriophorum vaginatum.* These 'islands' are raised above the general surface due to the caespitose growth of *E. vaginatum* and are roughly circular, of some 5 m in diameter. (They may perhaps be compared with the more humified surfaces found in, for example, peat section AII). Associated with the high water table community of *E. angustifolium* one has a differentiation based on the microrelief of the undulating *Sphagnum* ground layer. Mounds of *S. magellanicum* and *S. rubellum* support *Calluna vulgaris* and some *Eriophorum vaginatum,* while the slightly lower intervening areas of *Sphagnum cuspidatum* are richer in *Erica tetralix, Narthecium ossifragum* and *Rhynchospora alba,* as well as the ubiquitous *E. angustifolium.* Other species are very frequent but less noticeable in the vegetation because of their trailing habit *(Vaccinium oxycoccus)* or their sheltered position amongst the stems of higher plants *(Drosera rotundifolia).* A fairly high frequency scatter of *Andromeda polifolia* is a notable feature of this undisturbed vegetation and in a small area to the immediate south-west of the central copse there are a few plants of *Phragmites communis.* These show very poor growth (30-50 cm average) through carpets of mixed Sphagna with a peat depth of 2.6 m; they are comparable with the same species found at Coom Rigg Moss by Chapman (1964a).

An unpublished survey of the vegetation of the southern area of Bolton Fell Moss was made by D.A. Ratcliffe in January 1957 and has kindly been made available by the Nature Conservancy Council. The following list from this survey accords well with the author's field notes except in the minor variations noted below.

Eriophorum angustifolium	VA (very abundant)
Calluna vulgaris	VA
Erica tetralix	A (abundant)
[1]Andromeda polifolia	A
[2]Eriophorum vaginatum	F (frequent)
Oxycoccus palustris (now Vaccinium oxycoccus)	A
Narthecium ossifragum	F
Trichophorum cespitosum	F
Rhynchospora alba	F
Drosera rotundifolia	F
Sphagnum magellanicum	D (dominant)
[3]S. rubellum S. nemoreum (S. capillaceum)	VA
S. tenellum	VA
[4]S. cuspidatum	O (occasional)
S. papillosum	O
S. recurvum	O
[5]S. fuscum	R (rare)
Polytrichum alpestre	O
Odontoschisma sphagni	A
Leptoscyphus anomalus (Mylia anomala)	A
Lepidozia setacea	F

1. This species now seems to be somewhat reduced, perhaps due to burning. The author's field notes record it as Frequent.
2. Now rather more than Frequent, again perhaps due to burning.
3. Not easily separable in the field. Laboratory determinations indicate *S. rubellum.*
4. This species is at least Frequent and often locally Abundant.
5. *S. fuscum* not found by the author.

Within this vegetation type pools are lacking except after heavy rain, despite the very high water table – usually just below the surface. There is, however, plentiful evidence in small pits and short borings, of pool muds – greasy, yellow-green bands of *S. cuspidatum* and algae – at 10 cm or so below the surface. This accords very well with the stratigraphical sections (e.g. section HI) and it seems as though this undamaged surface is, as a whole, at a *Sphagnum* 'lawn' stage where the pools which once studded the surface have filled in, but before a possible stage of widespread hummock formation. This will be commented on further in discussion.

On the lines of the old drainage ditches which bound the south-west quadrant an interesting type of community has developed, promoted, it seems, by slightly better aeration and nutrient flow. Birch trees *(Betula pendula)* have colonised these areas (plate IV, and aerial photographs) to form a very open scrub of saplings and young trees up to 3 m high. Trees over this height show distinct signs of 'die-back' which seems to be related to sinking in the soft peat and death of the root system. A tree which had been pulled up during drainage operations in the south-east quadrant exhibited a root system which terminated in an abrupt line at about 30 cm depth, the roots turning upward from this level which presumably recorded the past depth of the aerobic horizon. Under the birches the accumulation of leaf-litter is accompanied by an abrupt change in the ground flora – less ombrotrophic Sphagna such as *Sphagnum fimbriatum S. palustre* and *S. recurvum* appear, and the Ericaceae are reduced. In the open areas between trees, *S. magellanicum* is still plentiful but other mosses are represented such as *Pleurozium schreberi, Acrocladium stramineum* (an indicator of 'slight mineral enrichment' and some water flow according to Watson 1968) and large mounds, up to 1 m across, of *Aulacomnium palustre* and *Polytrichum commune.*

The vegetation of the cut-over areas of the moss is dominated by luxuriant *Calluna vulgaris* and few associates, though *Eriophorum vaginatum* is locally dominant and *Erica tetralix* is frequent in slightly wetter areas (plate V). Birch saplings are frequent but are removed by the peat workers. Purple moor-grass *(Molinia caerulea)* is generally rare, as can be seen from the photographs, except on the bog edges – a point of some importance in the interpretation of the pollen diagrams.

The central copse of trees on boulder-clay is dominated by pine and birch with smaller amounts of oak *(Quercus petraea)* and some holly and ash. There is a rich ground flora of grasses, bracken. foxgloves and brambles. The present composition of the copse is thought to date from the reclamation attempt of 1800 AD – on the Record Office map (figure 12) only a single birch tree is shown. Species other than birch, with its extremely light seeds, would have difficulty colonizing such an isolated site and it seems probable that there was a birch wood on the site (without pine which was re-introduced in the early 19th century), and this was cleared in 1800 AD with the exception of a single 'marker' tree. The pollen evidence is certainly not at variance with this.

In summary it seems reasonable to conclude that the vegetation and stratigraphy of Bolton Fell Moss show beyond doubt that it is a site suitable for the testing of ideas on ombrotrophic bog growth.

3.2 OTHER CUMBRIAN SITES

When this study was first envisaged it was thought that comparative work on other bogs in the Cumbrian Lowland might form part of it. One of the reasons behind this was that as a rainfall gradient exists between the coast (32.5 inches per annum; 825 mm) and the higher area of the Cumbria/Northumberland border (45 inches per annum, 1 150 mm), then if suitable sites could be found some assessment of the importance of rainfall variation might be attempted. This was never thought of as a crucial part of the study, both for reasons of the time a detailed study of three or more sites would have taken and because, logically, falsification of the 'Osvald Theory' from a single site is sufficient. Nevertheless, it was thought that observations from more than one site would prove useful, if only to increase the experience of the author. In the event, work on sites other than Bolton Fell Moss was limited to observations rather than detailed analyses for the practical reasons mentioned below.

3.2.1 Butterburn Flow

This is an extensive (almost 4 km^2) area of bog surrounded by the 900 ft (275 m) contour at grid reference 670760. There are two fairly distinct areas of vegetation, both seemingly undamaged by drainage or cutting. South of the Lawrence Burn, which flows west-east into the River Irthing, the bog is dominated by very closely-spaced hummocks averaging 40 cm in height, roughly circular and about 30 cm in diameter, separated by narrow 20 cm channels rather than isodiametric hollows (plate VI). The water table is very high and the upper peat extremely sloppy, both in the hummocks and channels. The vegetation is very similar to that described from Coom Rigg Moss, only 3 km to the north, by Chapman (1964a), being dominated by *Eriophorum angustifolium* and *Sphagnum magellanicum. Erica tetralix* is almost invariably present on the tops of the hummocks with lesser quantities of *Calluna vulgaris* and *Sphagnum tenellum* dominating the channels, together with *S. cuspidatum.*

The northern part of Butterburn Flow is rather different, being an only slightly undulating *Sphagnum* lawn. *Sphagnum imbricatum* is present as low mounds and there are occasional isolated hummocks composed almost entirely of *Sphagnum fuscum. Sphagnum magellanicum* and *S. papillosum* are abundant under a cover of *Eriophorum angustifolium* but the dominance of this species appears reduced compared with the southern area – there are patches of uncovered Sphagna. In the central area (GR 675767) there are several large circular pools, some up to 2 m across with 30 cm of water above an extremely soft base. *Carex limosa* is present around these pools, as are a few plants of *Molinia caerulea.* Peat depth, ascertained by several boreholes, averaged 4-5 m.

It was hoped that the uppermost stratigraphy of Butterburn Flow could be investigated using a closely spaced grid of short cores. This proved to be impossible using a Hiller-type sampler because of the very unconsolidated nature of the peat and even with a Russian-type corer several attempts were necessary to extract even a fragmentary core. In most cores unhumified peat (H3) of

Sphagnum-Eriophorum occupied the top 20 cm, overlying a much more humified peat (H7-8) with obvious *Calluna* remains down to 100 cm. Greenish-yellow pool muds were visible at various depths and the stratigraphy is therefore not unlike that from Bolton Fell Moss. However, it was found impossible to correlate these borehole records with any certainty, even when the cores were only 20-25 cm apart. Correlations of pool muds could be made over distances of up to 80-100 cm but the stratigraphic status of bands of humified peat was much more difficult to ascertain. The reason for this is obvious when one inspects the stratigraphic section records of Bolton Fell Moss wherein hummocks are observed to expand and contract, to arise suddenly from a 'lawn'-type situation and to grow diagonally over pool layers (for example in section HI). These uncertainties were so great that it was felt that this work should not be proceeded with.

3.2.2 Glasson Moss and Wedholme Flow

As stated above it was hoped that some work on bogs at either end of the rainfall gradient across Cumbria might produce useful results and accordingly Glasson Moss (GR NY245605) was thought of as a possible site – average annual rainfall here is 32.5 inches (825 mm). Interest in this site also arose from the statement in Ratcliffe & Walker (1958) to the effect that whereas most Solway bogs show pool and hummock systems which, from casual observations, appear to remain stable through some depth of peat, Glasson Moss '. . . shows a series of lenses which demonstrate a process of bog growth in the classical manner of the regeneration complex demonstrated by Osvald in many other sites' (p.435). This statement is not enlarged upon in Walker (1966), a mainly pollen-analytical study. A further reason prompting some consideration of Glasson Moss was the knowledge that a local peat litter firm had cut and drained part of the bog.

Two visits were made to the site to assess the possibilities. It was found that the peat ditches were on the whole very shallow, averaging 30 cm, and this militated against comparative work with Bolton Fell Moss. (It has since been discovered that the peat company, finding a 'Grenz'-like layer of very humified peat near the surface assumed this to be the base of the *Sphagnum* peat and so abandoned the site. In fact some 5 m of *Sphagnum* peat lie beneath this humified layer.) However, more important than this was the fact that most of the moss had just then (1968) been severely damaged by a major fire, which following on an earlier fire in 1966, had burned and charred the peat to a depth of several centimetres. Despite a number of attempts at cutting back the peat faces no trace of a lenticular structure could be found. The stratigraphy appeared instead to equate with that of upper section H at Bolton Fell Moss with a series of shallow pools or hollows penetrated by small bosses of hummock peat. This accords well with observations of the uncut central area of

Plate I. Peat section at Kilberry Bog, Co. Kildare, Eire. The measuring rod is one metre long.

Plate II. Bolton Fell Moss: an aerial view of the northern part of the moss in 1951. The d
marks the approximate position of section HI (the vertical lines across this plate are artifacts produced by joining two separate aerial photographs).

Plate III. Bolton Fell Moss: an aerial view of the whole moss, 1974.

Plate IV. Bolton Fell Moss: a view of the southern part of the bog showing some of the original, uncut bog surface and the shallow new cuttings around Area B.

Plate V. Bolton Fell Moss: north-east quadrant showing area cut in 1800 AD and recut from 1957 on. Note the overgrown nature of the ditches and the mouldered appearance of the upper peat.

Plate VI. Butterburn Flow; showing extremely hummocky surface with channels surrounding hummocks.

Plate VII. Close-up of peat types at Bolton Fell Moss. The lower peat is of humification value 5 and contains obvious *Calluna* remains. The upper *Sphagnum* peat, of humification 3, is almost pure; the layering may be due to periods of drier summers.

the moss where shallow pools, containing free water only after heavy rain (depth c.5-10 cm) and with a vegetation of *Sphagnum cuspidatum* and *Rhynchospora alba*, alternate with low hummocks capped by *S. rubellum, Calluna vulgaris* and *Erica tetralix*.

Subsequent correspondence with Professor Walker confirmed these observations and he now doubts whether the type of stratigraphy which he noted from Glasson Moss was truly lenticular. The moss has now been more or less destroyed by a fire during the 1976 drought which burned for four weeks (D. A. Ratcliffe, personal communication).

Wedholme Flow (GR NY220530) was also investigated as a possible alternative site. Peat-cutting is very active over the whole of this bog but the actual mode of operation was found to be very different to that obtaining at Bolton Fell Moss. A number of large deep ditches had been dug across the bog at the start of operations in the 1950's; these were very much overgrown and the sides had fallen in making them unsuitable for detailed stratigraphic work. Following this the whole bog had been cut-over to one spade depth (c.20 cm), removing almost the whole of the uppermost stratigraphy by 1967; a further cycle of work to remove the next 20 cm or so of peat had already begun by that date. It was obvious therefore that this site also was unsuitable for further investigation.

However, observations of a small area of uncut bog remaining near the centre of Wedholme Flow – a Site of Special Scientific Interest notified by the Nature Conservancy Council – showed a vegetation type more or less identical to the uncut south-west quadrant of Bolton Fell Moss, also an SSSI. Similarly an examination of a few remnants of the uppermost peat and of the ditches of the 'second cut' revealed a stratigraphy indistinguishable from that at Bolton Fell Moss so that this latter site may not unreasonably be taken to be typical of the Solway bogs as a whole.

3.3 OTHER SITES

It was recognised from the outset of this work that visits to other bogs in the British Isles and elsewhere would be valuable in increasing the author's knowledge of bog ecology and peat stratigraphy. Accordingly a number and variety of sites were visited and detailed field notes, sketches and photographs taken. Where relevant such observations are included in later sections of this work and the range of sites considered is set out briefly below.

Although it has not been possible to visit Swedish and Finnish bogs illustrated publications from these countries have been closely studied (the works of Osvald and Tolonen in particular contain many excellent photographs) and the bogs studied by Kulczynski (1949) are illustrated with 46 plates. Within the local area of Bolton Fell Moss a number of cut and uncut bogs were visited including Scaleby Moss, Solway Moss, and Bowness Common, besides the sites already mentioned. To the north the large raised bog of Flanders Moss, and numerous other mires were studied in the company of various European phytosociologists on the 1968 International Phytogeographical Excursion to Scotland. This study tour proved invaluable in giving the author an insight into Continental views on mire ecology.

The large raised bogs of Ireland were also visited in 1968 and it proved possible to visit most of the sites studied by the Walkers in 1959, including, it is thought, the same peat face at Ballymacombes More recorded by the Walkers (1961). The bogs at Fallahogy, Clonsast, Newbridge and Rathernin were also included and an interesting type of 'regenerative' stratigraphy (not seen by the Walkers) was noted at a bog near Kilberry, County Kildare (plate I).

Tregaron Bog, site of Godwin & Conway's classic work (1939) was closely studied in 1970, when direct evidence of old fires was found in the form of charred *Calluna* twigs at and just below the present surface, and Borth Bog, a much more *Sphagnum*-rich site with large pools, was also visited. More recently since the author came to live in Southern England, a variety of mires of varying trophic status have been worked upon, especially in the Hampshire-Dorset Tertiary basin. Whilst none of these individual experiences can be said to have decisively affected the work on Bolton Fell Moss, collectively they have helped the author to consider factors such as floristic and hydrological differences in a wider context.

4. METHODS

A detailed review of many of the methods used to investigate the history of vegetation has recently been published by the author (Barber 1976). It contains a number of references to standard guides for the identification of macrofossils, pollen and spores, rhizopods and other remains, including drawings of some specimens and a standardised method of pollen preparation. In view of this only those techniques strictly relevant to the present study are dealt with below and reference should be made to the above publication for information on wider applications of the methods.

4.1 PEAT STRATIGRAPHY

4.1.1 Site selection

The particular reasons for the choice of each site for stratigraphic recording are given in the next section (5. Results) but it is appropriate to note here the general reasons governing site selection. It was important that the peat sections recorded should be as representative as possible of the former bog-plane area, the area which could, in the past, have belonged to a Regeneration Complex *sensu* Osvald. Sections near to the original sloping marginal 'rand' of the bog were therefore excluded, as were those near an obvious mineral-ground water source, such as the boulder-clay inlier. Two further constraints lay in the need to generally avoid the north-east quadrant cut-over in the 19th century, and the preferability of using deep peat sections so as to extend the record back in time as far as possible. A strictly random approach was therefore inappropriate for these reasons, as well as because of the time-consuming nature of the laboratory analyses. An element of randomness was, however, introduced in for example Area A – the section recorded was determined by deciding to record the southern part of the fourth peat-face to the west of a major junction between the main drainage ditches. In determining the length of a recorded section the general principles adopted were that the length should be as representative as possible of the stratigraphy of the peat-face as a whole, but not less than 300 cm, and that recording should only terminate in stratigraphy which could be observed to be more or less similar for a further 100 cm or so. These guidelines did not of course apply in the case of the serial sections cut behind part of Section H, where the objective was to look at three-dimensional form of the particularly interesting central part of the recorded section.

On the basis of these principles, and on observation and field-sketching of several hundreds of metres of peat-face it is thought that the sections recorded in detail are a fair representation of the general stratigraphy and its main variations at Bolton Fell Moss, and, as has already been argued, Bolton Fell Moss is quite comparable with other ombrotrophic bogs of north-west Europe from which regeneration complexes have been described, especially with regard to species composition.

4.1.2 Conventions

Previous workers such as Walker & Walker (1961) and Casparie (1969) divided their peat types into six classes on the basis of gross macrofossil content and humification. This was thought to be undesirable in the present study on the

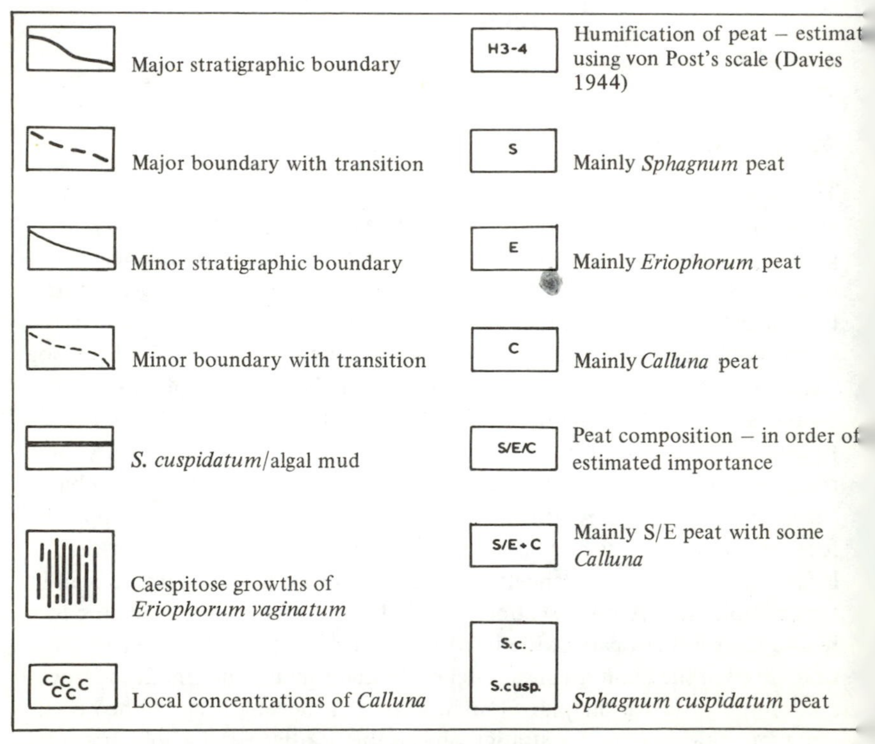

Figure 13. Key to peat stratigraphic symbols and conventions.

grounds that it may lead to loss of information by not adequately expressing the variation in the boundaries between different strata or the variation in content. The Troels-Smith system (1955) was tested and rejected for similar reasons, and so a new system of conventions was devised and is described below and in figure 13.

Boundaries between peat types were classified into four types instead of the single class used by previous authors. A major stratigraphic boundary, drawn as a heavy (1 mm) line, was defined as a boundary between peats of widely differing humification and composition – for example, between *Calluna-Sphagnum* peat of humification 6-7 (on von Post's scale, as given by Davies 1944) and fresh *Sphagnum* peat of humification 2-3. A minor boundary (line thickness 0.2 mm) was used to indicate lesser changes of humification and/or composition, between for example *Sphagnum-Calluna* H4-5 peat and *Sphagnum* H3 peat. Where the boundary line between one peat-type and another occupied one centimetre or less a solid line was used and a dashed line was used for transitional boundaries, both major and minor, where the peat-types graded into one another over a distance of more than 1 cm. This system reflects fairly accurately what could be discerned in the field and can be observed in the photographic record, although it must be admitted that such is the complexity of the field situation that not every boundary could be displayed absolutely accurately on the section diagrams.

Peat components were recorded using a combination of signatures. The striking bands of yellowish-green pool-mud, usually only 1-2 cm thick, were recorded as double lines 1 mm or so apart; local concentrations of caespitose *Eriophorum vaginatum* as vertical lines approximately to its extent in the field (a visually useful convention employed by both Walker and Casparie); and local concentrations of *Calluna* twigs, roots, etc., as capital letter 'C's. Peat of mixed composition – the major part of all sections – was classified according to the estimated importance of the components prefixed by the grade of humification. Hence 'H4S/E/C' denotes peat in which *Sphagnum* species were most important but with significant amounts of *Eriophorum* (both species) and slightly lesser amounts of *Calluna,* with an overall humification of grade 4; while 'H2-3S+C' denotes mainly *Sphagnum* peat of low humification, with a few noticeable *Calluna* remains. 'H3S' peat, a very common type of fresh, gingery-looking peat, may sometimes contain a few fibres of *Eriophorum,* or the odd small twig of *Calluna,* but these were difficult to detect and in this peat type, *Sphagnum* remains were always well over 90 % of the deposit. The close-up photograph (plate VII), showing 'H3S' over 'H5S/C' makes this clear.

Peat composed mainly of the remains of *Sphagnum cuspidatum* was recorded as a special type, denoted 'S.c.' or 'S.cusp.', as it could be separated from others in the field, being generally green, sometimes laminated or 'felted' and with leaves and sometimes full plants distinguishable by eye. It was also often associated with pool-muds ('p.m.') and on occasion with other *Sphagnum* species (probably *S. subsecundum*) in which case it was recorded as 'Pool Sphagna' ('Pool S.').

This system of conventions gives stratigraphic diagrams which may appear at first sight more complex than those of Walker and Casparie but which, it is thought, are in fact easier to read once the conventions are familiar, in that all the information is on the actual diagram and separate keys do not need to be consulted frequently. The diagrams also contain more information. Purely as a visual aid however, simplified stratigraphic diagrams are included in the Results, section 5.

4.1.3 Field recording

Once the section to be recorded had been determined a series of colour photographs (both slide and print film) of the peat-face were taken. This was seen as an important 'archive' record to be done prior to any disturbance by recording and sampling. 'Agfa' brand films were used, after a test using a number of brands, for their good ability to render the various shades of brown (CN17 print film; CT18 reversal film).

The stratigraphy was then recorded by pegging a 1 m^2 frame with 10 cm wire divisions to the peat face, and transferring the stratigraphic detail by pencil onto cm/mm graph paper. Wire frames are more suitable than string which 'snags' on *Calluna* twigs. Using this method gives an accuracy of ±0.5 cm (Walker & Walker 1961) and all peat-faces were checked more than once. Stratigraphy was, as far as possible, recorded from weathered peat-faces which show more detail, especially with regard to *Calluna* and *Eriophorum* remains, than a recently dug or 'cleaned-up' section. It was found necessary to leave the serial sections cut into the centre of section H for a minimum of three weeks to allow rain to wash away the loose peat smeared across the newly-cut face and for *Calluna* and *Eriophorum* remains to take on a more noticeable sun-bleached appearance.

Peat composition was estimated in the field from at least two samples from each peat-type. Generally a block of peat 5 x 5 x 5 cm was extracted, broken open and examined by eye and with x10 and x20 lenses. Representative samples of every type were also examined in the laboratory using a stereo-microscope and checked against the field record. Differences were found to be mainly related to the relative proportions of *Calluna* and *Eriophorum* in that very small twigs and leaf remains were under-recorded in the field record, and the same sort of small differences are apparent from the detailed macrofossil analyses (Results, section 5). These small differences are an inherent limitation of field recording and the general match of field records and macrofossil analyses is nevertheless a good one.

The von Post humification method was adopted as the most practical and informative method (Davies 1944, Appendix I). Other methods were considered and tested (e.g. Overbeck's (1957) alkaline extract method, humic material being estimated colorimetrically) but rejected as impossible for field use and time-consuming in the laboratory. Besides, the von Post method has the advantage of not only estimating humosity of the liquid contained in the peat but also of taking account of the degree of breakdown of the plant structures – that is, it expresses the totality of the humification process. The system is well defined and has been used by all former peat researchers. Humifications were determined on the same pieces of peat used to establish composition and the frequent use of intermediate values (H2-3; H5-6, etc.), on the stratigraphic diagrams expresses the slight variability found in a number of samples of what was obviously the same stratigraphic unit.

Two particular types of peat presented problems with regard to humification – pool peat and the uppermost peat in some sections. The former type, mainly *Sphagnum cuspidatum* and *S. subsecundum,* was often 'greasy' due to algal remains and in humification tests in the field would almost all squeeze out of one's hand – characteristic of highly-humified H9 peat. In the same sample however the plant structures were only a little obscured (H5 approximately) and the expressed water was only slightly turbid, characteristic of H2-3. In these cases therefore either no humification figure was recorded (in very greasy peat) or a compromise value of H5 or so decided on. The uppermost 5-10 cm of peat in some sections presented a rather different problem in that it could be seen to be comprised of both more or less unhumified (H1-2) *Sphagnum* remains (clear, pale-yellow water on squeezing) and rather humified, blackened twigs of *Calluna.* in some cases a proportion of more humified *Sphagnum* remains were also present, probably representing the different rates of breakdown of two or more species. Again some sort of compromise figure was recorded, based on the proportions of the various plant components.

4.1.4. Monolith sampling

Once the stratigraphic sections had been drawn up and re-checked in the field, sampling for macrofossil, pollen and radiocarbon analyses could be considered. The particular reasons for choosing each profile sample are given in section 5 but the general principles followed were that sampling should be representative of both hummock and hollow environments and should be as complete as possible bearing in mind the constraints of time available for analyses and the availability of monolith tins – these, being made of aluminium, were expensive. In all 45 tins were used, each 50 x 10 x 10 cm, of aluminium sheet 1.5 mm thick; virtually no corrosion of the aluminium has taken place after 10 years contact with wet, acidic peat.

The labelled tins were pressed into the peat then removed and the outline so formed was cut around with a sharp knife to sever *Eriophorum* fibres and the like. The tins were then lined with aluminium foil to facilitate removal of the peat, so that the face actually recorded could be studied and subsampled rather than the peat 10 cm behind the face, and the tins were then pressed home with as little disturbance as possible, removed by cutting behind them with a peat spade and wrapped in aluminium foil and polythene sheet. Stratigraphy was checked again immediately on return to the laboratory and the monoliths were then stored tightly sealed at room temperature or in a freezer. Possible distortion of the stratigraphy due to freezing and thawing was checked experimentally and found to be undetectable.

The advantage of such large monoliths was that the same sample could be sub-divided for all subsequent analyses, including radiocarbon date samples, whereas Casparie (1972) used iron tins (30 x 2.5 x 3 cm) and 'spot' pieces of peat cut out and packed into polythene bags, which lack continuity and increase the possibility of mistaken correlations and contamination.

4.2 MACROFOSSIL ANALYSES

While a great deal of information on peat accumulation may be gained from the study of stratigraphic sections alone, or sections wherein the strata are correlated by pollen-analysis and dated by radiocarbon assay, much more evidence can be brought to bear on the problem by analysis of the macrofossils that make up the peat. This was recognised by Walker & Walker (1961) who analysed three monoliths for macrofossils and built upon by Tolonen (1966, 1971) and Casparie (1969, 1972) both of whom used macrofossil analysis extensively. At Bolton Fell Moss some macrofossil analyses were performed at an early stage of the study using chemical maceration techniques but this proved unsatisfactory and all the analyses were done later, after pollen correlation of the stratigraphy, using the standardised method described below.

4.2.1 Sampling method

Each monolith was turned out of its tin and the face recorded in the field was re-examined using an illuminated magnifier. The monolith stratigraphy was recorded independently of the results of the section stratigraphy and, not surprisingly, some slight variants in the stratigraphy which had not been noticed in the field were found, though the overall correspondence between field and laboratory observations was very good.

Sampling at regular intervals was not considered appropriate; instead the principle adopted was that each stratum visible should be sampled at least once. This was adhered to for all monoliths from peat sections A to G and for the main monoliths from section H, but relaxed in the case of the less important monoliths from that section which were used for filling-in various details. Constant volume samples, 5 x 2 x 1 cm thick, were cut from the monolith using two sharpened steel plates and stored in petri dishes at a low temperature until treatment.

4.2.2 Treatment and assessment

The usual methods of treating macrofossil remains – maceration in dilute caustic soda or nitric acid (Godwin 1956) – were found to be quite unnecessary in dealing with the Bolton Fell Moss material. Not only do these chemicals have disadvantages in rendering the remains mucilaginous (dil.NaOH), in requiring extra care in handling and in reacting with the mounting media, but the reasons for using them – solution of humic colloids and disaggregation of the sediment – were found to lead to a loss of information for the purposes of this study. Instead of any chemicals plain tap water was used to wet sieve the samples in a semi-quantitative way. In the first series of analyses a nest of five 8 inch sieves was used – 2000, 1000, 500, 250 and 125 microns mesh size. This was also found to be unnecessary; though it may be useful in heterogenous deposits where wood, leaves and seeds of different sizes are found, in *Sphagnum* peats it merely reflected the degree of breakdown, both natural and induced, of the material. The first three sieve sizes having been abandoned for this reason it was found that no identifiable *Sphagnum* leaf fragments were retained by the 125 micron sieve and so only the 250 micron size was used.

By placing this sieve on a three litre beaker the 10 cc sample could be washed with a standard amount of water to fill the beaker. The colour of the washing water, from almost completely clear to a turbid black, was then used to estimate the amount of 'unidentified organic matter' on a five-point scale; this correlated closely with the humification values. All the material remaining on the sieve was then transferred in water to an oblong glass trough sufficiently large to give a spread of material observable with transmitted light. This material was systematically scanned using a Nikon stereozoom microscope (SMZ-2) of stepless x8-x40 magnification and assessed on a five-point scale, as used by Walker & Walker (1961), and other workers, where score 1 = rare, 2 = occasional, 3 = frequent, 4 = common, and 5 = abundant. Although subjective this assessment procedure proved to be replicatable when the same samples were re-assessed after an interval of several weeks. Certainly it is difficult to see what other method could be used with this sort of material. Green (1965) counted the actual number of branch leaves of various groups of Sphagna in glycerine jelly slide preparations and presented results in percentage diagrams rather like pollen diagrams. This method is suspect, however, as Green realized (1965, pp.102-105), due not only to identification problems on branch leaves alone but also because the Acutifolia group of Sphagna have small very abundant leaves compared with the Cymbifolia group, and moreover, it is the rule to find single leaves of the former group whereas in Cymbifolian species whole shoots and branches remain intact. No published work has yet used this method, which also, of course, excludes the large remains of *Calluna* and *Eriophorum.** Heikurainen & Huikari (1952) introduced a point frequency method involving 100 point determinations of peat constituents per slide but this is again based on the dubious method of identifying branch leaves by their cell structure, as well as being open to the other criticisms outlined above, and has not been adopted by other workers.

Following the trough estimation stage a small spatula was used to extract three random collections of the Sphagna present – these were mixed and placed on a microscope slide, a gum chloral type mountant added (Berleses' Fluid was found to give more durable slides than Farrant's Medium) sufficient to give a self-sealing mount under a 40 x 20 mm cover slip. Most slides prepared in this way contained at least 50 leaves and included whole branches of Cymbifolian species. Total Sphagna, which had been estimated in the trough sample, was thus broken down into the species and groups used to construct the macrofossil diagrams by examining all leaves on a slide at magnifications of x25 to x400 using a Nikon SKt compound microscope with widefield condenser. Where the proportions estimated by this method seemed at variance with the preliminary estimates made at the trough stage (in, for example, proportions of *S. cuspidatum* leaves to leaves of Cymbifolian species) a further slide was prepared and counted. Such variations were not common, many slides containing only two species.

4.2.3 Identification

The identification of sub-fossil Sphagna presents a number of problems but quite fortuitously most of the important ombrotrophic species are readily determinable even from single leaves. The genus *Sphagnum* comprises two subgenera, totalling 43 species in Europe generally (Isoviita 1966) of which 30 species are found in the British Isles (Dickson 1973). The cymbifolian ('boat-shaped', cucullate) leaved species of the sub-genus *Inophloea* 'includes the most important peat-forming Sphagna' (Proctor 1955) and it is these five species which are easy to recognize, both living and sub-fossil. Of these five, *Sphagnum palustre* is generally found in more base-rich habitats than raised-bog being very widespread in wet woodlands, soligenous mires and stream sides (Dickson 1973). No specimens were found in the Bolton Fell Moss peat. *S. centrale* is not mentioned by Dickson (1973) and is equated with *S. palustre* by Duncan (1962) while Proctor (1955) doubts its taxonomic status but records it as 'widely scattered but rare, with a generally northern distribution'. Richards & Wallace (1950) – whose nomenclature is used here rather than the well-founded but not widely known revision by Isoviita (1966) – record it as of

* Green has recently published his 1965 leaf counts, with modifications and notes on their possible inaccuracies – *J. Ecol.* 1977, v.65, pp.797-9.

full specific rank and Isoviita (1966) also regards it is of 'indisputable specific status'. This interesting species was therefore searched for carefully during the analyses but without success. We are left therefore with only three cymbifolian species – *S. imbricatum,* easily distinguished by the comb fibrils of the hyaline cells; *S. papillosum,* equally easily identified by the papillose walls of the hyaline cells; and *S. magellanicum* wherein, by careful differential focussing at x100 or more, the chlorophyllose cells may be seen to be enclosed by the hyaline cells on both leaf surfaces, as illustrated especially well by Fearnsides (1938). It was further discovered, during the course of the analyses, that the presence of *S. imbricatum* could be successfully predicted at the trough stage by the high degree of integrity of branches and stem fragments.

As these three species form the great bulk of the peat at Bolton Fell Moss it is encouraging that their identification is so certain, for the other sub-genus, *Litophloea,* 25 species in all, does involve difficult identification problems. These are partially offset by the fact that many of the species occupy similar habitats and that one particular species, *S. cuspidatum,* is readily determined by the great length of its branch leaves (4-5 times as long as broad according to Proctor (1955)). Furthermore quote a few species may be ruled out on grounds of rarity – *S. obtusum, S. lindbergii, S. riparium, S. balticum, S. quinquefarium, S. molle* and *S. warnstorfianum.* This may seem dangerous practice but a careful survey of the literature concerning present distribution and the fossil record (Dickson 1973) is felt to justify this step, and in any case these species are not separable in the usual fossil state. A further large group may reasonably be excluded from consideration on ecological grounds, both generally, from the literature and the author's observations of various mires, and by particular consideration of the present flora of Bolton Fell Moss. This group includes species which are found in fen woods and areas of relatively high base-status – *S. fimbriatum, S. girgensohnii, S. squarrosum, S. contortum* and *S. teres,* and two species which are confined to damp heathland and moor rather than raised bog – *S. compactum* and *S. strictum. S. pulchrum* can also probably be disregarded on distributional grounds, the literature referring to it as a species of 'poor fen (and sometimes bog) especially in the south' (Proctor 1955).

The eight remaining species may be briefly dealt with individually. Three species belong to the *Cuspidata* section, *Sphagnum tenellum, S. recurvum* and *S. cuspidatum.* The latter species is easily recognizable on two characters (Fearnsides 1938) – its triangular fibrillate stem leaves and its very long branch leaves; indeed these typical, often sickle-shaped leaves are quite obvious in the trough sample. *S. recurvum* and *S. tenellum* present a problem of separation when, as in the sub-fossil situation, their stem and branch leaves are met with singly. In live specimens they present no difficulty to the experienced bryologist using characteristics such as the fibrils in the upper third of the stem leaves of *S. tenellum* and in the field this species characteristically grows as single plants scattered amongst other Sphagna, but there are only five fossil records (Dickson 1973), two of these from spores alone (Tallis 1964). There are no fossil records of *S. recurvum* (Dickson 1973). Once more, however, there are good ecological reasons for doubting their importance, if not their presence, in the Bolton Fell Moss peat. As noted above, *S. tenellum* occurs as single plants and though present in the Sphagnetum of Bolton Fell Moss it can in no way be regarded as an important peat-former. *S. recurvum,* which is occasional in the present vegetation of the bog is nevertheless only noticeable in areas such as the scattered birches and depauperate *Phragmites* of the south-west quadrant according to the author's field notes, and this is in agreement with Proctor's (1955) verdict: '. . . usually in somewhat eutrophic conditions, though not as base-tolerant as *S. squarrosum;* often forming extensive carpets in poor fen'. Certainly it is not regarded as a species of the Regeneration Complex or of raised-bog vegetation generally except near the edges, or in ditches in association with *Molinia caerulea* and some soil water influence (for example, as recorded by Chapman (1964a)). For all these reasons then, *S. tenellum* and *S. recurvum* are not recorded in the macrofossil analyses.

Sphagnum subsecundum occurs in pools, generally at a lower level than the common *S. cuspidatum.* Some confusion exists in the literature regarding the taxonomic status of the species.* Most British bryologists regard it to have a number of varieties: var. *inundatum,* var. *auriculatum,* var. *bavaricum* and var. *rufescens,* according to Richards & Wallace (1950), the latter two being rare (Proctor 1955). Continental workers recognise the first two varieties as full species together with *S. subsecundum* itself (Isoviita 1966). Whatever its status the only readily determinable fossil leaves are the large robust ones of var. *inundatum* and var. *auriculatum,* both more common than the 'parent' species, with var. *inundatum* the more likely in acid habitats according to Duncan (1962). Watson (1968) notes that var. *auriculatum* occurs especially in 'poor fen' and other areas which some mineral enrichment whereas var. *inundatum* 'often grows with *S. cuspidatum* submerged in deep pools'. The balance of opinion therefore favours var. *inundatum* as the most probable plant found in the fossil pools – as no pools now occur on the bog no check with living material was possible – but whichever variety was present makes little difference to the ecological conclusions drawn from the results, and in view of these uncertainties and the taxonomic position the author decided to err on the side of caution, recording it simply as *S. subsecundum.*

The final four species all belong to the Acutifolia section – *Sphagnum rubellum, S. nemoreum, S. fuscum* and *S. plumulosum.* Field evidence is against the widespread occurrence of *S. plumulosum* ('somewhat base-rich . . . often with much *Molinia'*, Proctor (1955)) on Bolton Fell Moss, as it is against *S. fuscum,* a readily identifiable brown, densely-branched species. Fossil evidence of branch leaf size also leads one to conclude that *S. plumulosum,* with

* Status of species recently revised by Eddy (1977) – *J. Bryol.* 9:309-319.

leaves half as large again as the other species, is not important in the peat. On the other hand both *S. rubellum* and *S. nemoreum* are abundant in the present vegetation of Bolton Fell Moss, are difficult to separate in the field (Watson 1968, Chapman 1964a) and can not be distinguished with any confidence at all in their usual sub-fossil state of abundant single leaves. The author's impression of their relative abundance in the field is that we are dealing mainly with *S. rubellum,* with its always characteristic deep red hue and its 'delicateness' compared with *S. nemoreum.* In a number of fresh collections examined microscopically *S. rubellum* also predominates. The evidence is therefore quite good for the view that more or less all the Acutifolia identified belong to these two species but in any case all four species are hummock-builders – that is, of the same ecological group – and they are aggregated on the macrofossil diagrams as 'Sphagna Acutifolia'.

Needless to say Sphagna identification were not made on the use of keys alone; as C.A. Dickson (1970) warns: 'It can be deceptively easy to key out a species with the aid of a handbook alone; gross errors have resulted from such action'. Mindful of this the author prepared reference slides of all the common Sphagna, both from fresh specimens gathered from Bolton Fell Moss and various other bogs, and from authenticated herbarium material kindly supplied by Professor C.D. Pigott. Two sets were made, the first using a modified form of the laborious preparation technique of Fearnsides (1938), which were stained; the second was simply made up in Berlese's Fluid and not stained – these were more comparable with the sub-fossil material. Such slides were referred to constantly in the early stages of the work and, it must be noted, special efforts were made to find 'unexpected' sub-fossil species – the species 'ruled out' of consideration in the above account were only done so when the author had become experienced in 'sphagnology'. In addition all 30 species were included in a card index made up with the use of Fearnsides (1938), Proctor (1955), Duncan (1962), Isoviita (1966) and other sources.

Finally, mention should be made of the identification of macrofossil remains other than Sphagna. Considering the variety of the present vegetation of the bog I was surprised at the general lack of remains of other mosses and of vascular plants other than *Calluna* and *Eriophorum.* No other genera of mosses were encountered, including the very characteristic leaves of *Polytrichum.* This does, however, seem to be a general experience – none were recorded by Walker & Walker (1961), and Casparie (1969) notes them, at three levels only, below the upper Sphagnaceous peat – and is probably explained by its very local distribution on the bog today. A lot of effort was expended in the early analyses in attempting to identify a wide range of fragments of vascular plants – *Rhynchospora alba, Narthecium ossifragum, Trichophorum caespitosum,* etc. – which must have been present in small quantities but these lack specific identifiable characteristics (fresh material was, of course, examined) and the odd well-preserved specimen has been omitted from the diagrams for the sake of clarity. Likewise small twigs, leaves and even flowers of *Erica tetralix, Andromeda polifolia* and *Vaccinium oxycoccus* were encountered and included in the first diagrams as a scatter of scale 1 and 2 values but subsequently omitted as not repaying the effort involved – Walker & Walker (1961) seem to have reached the same conclusion.

Eriophorum angustifolium deserves a special mention as a dominant in the present vegetation but not represented in the macrofossil diagrams – nor is it in the diagrams of Walker & Walker (1961) or Casparie (1969, 1972). This is because whereas *E. vaginatum* occurs as densely- tufted masses with blackened sclerenchymatous spindles from the leaf-bases (C.A. Dickson 1970), *E. angustifolium* occurs as scattered, single 'adventitious roots of a fresh, pink colour' (Godwin 1975) which seem to decay more rapidly. Its rhizomatous habit, producing separate aerial shoots, also means that the level at which the remains are found are not a reliable indicator of the former ground surface (Godwin 1975) so while it was found in the *Sphagnum cuspidatum* peats as expected, I have followed previous authors in omitting it from the diagrams.

From all the above considerations it will be seen that the picture which emerges is essentially one of interactions between a number of important peat-forming Sphagna, *Calluna vulgaris* and *Eriophorum vaginatum,* the essential elements of the Regeneration Complex.

4.2.4 Method of presentation of results

Bar-graphs are considered to be the clearest and most objective method of presentation in that the actual values at each level may be read off the graph. The width of the bars is at a true scale – that is, each sample is shown as a 1 cm bar related to the depth scale. The advantage of 'readability' of a 'resolved' (blacked-in) diagram was also thought important and so a light toning ('Letratone') was applied to help visualization of the trends. In the interests of objectivity this toning was only used between two adjacent values and not brought down to the zero-line at an arbitrary point – it is not suggested that the 'real' values of intermediate levels lie on a straight line between two adjacent values. In a few cases the uppermost value of *Sphagnum imbricatum* is presented as an open bar to indicate the probability of contamination by cut peat blocks left to dry on the surface.

The arrangement of the diagram components follows a scheme loosely based on the species relationships to the water level – from *Sphagnum subsecundum* on the left to *Calluna* and *Eriophorum* on the right – and on the apparent 'successional' relationships between *Sphagnum imbricatum, S. papillosum* and *S. magellanicum,* with other components (Total Sphagna, Unidentified Organic Matter) at either end. Information on humification, stratigraphic boundaries and peat type, as recorded from laboratory examination of the monoliths is given in a column to the left, with radiocarbon dates where available. As an aid to the interpretation of each diagram and to save frequent referral to the stratigraphic section diagrams, a brief description of the stratigraphic situation is given to the left of the peat type column.

4.3 POLLEN ANALYSES

The main purpose of pollen-analysing the peat from Bolton Fell Moss was to provide a framework for the correlation of the stratigraphy. To achieve this pollen diagrams were constructed from most of the monoliths, special attention being paid to stratigraphic boundaries. The first two pollen diagrams, from sections A and B, were important in establishing the rather different chronology of these two sections due to human interference with drainage near section A and thus alerted the author to the possibility that section C was not all that it seemed. Pollen analyses from section C later confirmed that this section had been disturbed, with grievous results for the progress of the study, so that work on the macrofossils and radiocarbon-dating of the main section, section H, was preceded by thorough pollen analytical work on the uppermost peat to check that it was intact. This led to the choice of a particular monolith from section H, monolith 9, for the construction of a close-counted 'master' pollen diagram with which all other diagrams have been correlated. The preparation of this diagram took some months but it has proved most useful in a number of ways other than providing the correlative framework for the stratigraphy. For example, it proved possible to relate the pollen diagram to the documented land-use history of the area – as pioneered by Oldfield (1960, 1963) – thus providing a much finer chronology than that possible by the use of radiocarbon-dating alone. Furthermore, the diagram itself has important 'spin-off' value as documentation of the vegetational, or rather agricultural, history of northern Cumbria since pre-Roman times.

The interpretation of this diagram, and the other less detailed profiles, is considered in section 5.4; the methods used in their construction are described below.

4.3.1 Sampling

All samples for pollen-analysis were taken from freshly-exposed surfaces of monoliths or Russian-type sampler cores (Barber 1976, section 2.1.3). One centimetre thick lumps of peat (0.5 cm either side of the mark) were extracted with a pair of spatulas, exercising the greatest care to avoid contamination – for example, master profile HI9 was sampled in winter and in a 'clean room' with force-filtered ventilation to minimise air-borne modern pollen contamination. Samples were stored in closed vials in a refrigerator until needed – a standard 1 cm^3 was used in the chemical preparation.

Sampling intervals were determined by the needs of the work, using a basic 2 cm interval for the master diagram with contiguous 1 cm samples used for especially critical horizons. Elsewhere a basic interval of 5 cm was used (e.g. section C) or close samples spaced across stratigraphic boundaries (e.g. monolith HI6).

4.3.2 Extraction of pollen and spores

After experimentation with various published and unpublished preparation techniques (Faegri & Iversen 1964, Brown 1960, Gray 1965, West 1968 and Lancaster University methods) a standardised schedule was adopted which the author has since published (Barber 1976). It was not necessary to follow this schedule in full for pollen extraction from the uppermost peats – no carbonates were present so that HCl treatment was unnecessary as was HF treatment. Conversely the high humification of some of the hummock peats was found to require a double treatment, including sieving, with caustic alkali. In some hummock samples a very high proportion of finely divided lignaceous material was found and oxidation using sodium chlorate (Faegri & Iversen 1964) was considered in an attempt to remove this. These were abandoned after it was found that the preparation time needed to make any real impression on the amount of lignin present was having a deleterious affect on the pollen grains.

Safranin 0 was generally used as a stain but supplanted in later work by Bismarck Brown which has little tendency to overstain grains thus helping in critical identification (Brown 1960). Silicone fluid (MS200/2000 centistokes viscosity) was used throughout as a mounting medium, being superior to the commonly used glycerol or glycerine jelly. The earliest preparations mounted in silicone fluid show no detectable change after ten years; an archive of all vials and slides in this mountant has been established.

4.3.3 Identification and counting

Since the Bolton Fell Moss samples included many samples rich in agricultural weed species, various specialized keys and monographs were consulted as well as a full pollen reference collection. The basic text used was Faegri & Iversen (1964), which includes the standard key to North-West European pollen types, supplemented by the pollen atlases of Erdtman (1943), Erdtman, Berglund & Praglowski (1961) and Erdtman, Praglowski & Nilsson (1963), as well as by Godwin (1956). Special attention was paid to a number of critical species, notably the grains of cereals, of *Cannabis/Humulus* type, and of exotic trees – e.g. *Juglans regia.* Cereal-type pollens were differentiated using the criteria of Faegri & Iversen's (1964) special *Gramineae* key and Beug's (1961) criteria. Grains were examined using phase-contrast optics and all were measured for maximum diameter and annulus diameter in an attempt to separate *Triticum, Secale, Avena* and *Hordeum.* Although in early work the author felt that some well-preserved grains were separable on Beug's (1961) criteria, Andersen & Bertelsen (1972) have since shown these criteria to be suspect so that the Bolton Fell Moss results are expressed only as 'Cerealia' and *Secale.* Even so this distinction is useful as a chronological marker (Godwin 1975).

Pollen of *Cannabis/Humulus* type is also a useful chronological indicator, not so much by its first appearance as by its demise at about 1800 AD. Pollen of this type was examined closely, and on the basis of the criteria used by Godwin (1967a, b) and by comparison with reference material of the two species the author is certain of its identification as *Cannabis sativa.* This identification was confirmed by Miss Robin Andrew (Cambridge Botany School), as was the identification of a single grain of *Vitis vinifera,* the vine.

A number of grains of exotic tree pollen were also found, specifically *Picea* and *Juglans,* whose indicator value is discussed later. Pollen from the three species introduced pre-1800 AD (Mitchell 1974) – *Juglans regia* (Walnut), *Juglans nigra* (Black Walnut) and *Juglans cinerea* (Butter-Nut) – was kindly supplied by the Liverpool University Herbarium, to whom I am also indebted for specimens of *Cannabis* and *Humulus.* Comparison with these reference slides identified the fossil pollen to be *Juglans regia,* as expected.

Pollen recorded as *Corylus* is thought to be definitely *Corylus avellana* rather than the closely similar *Myrica gale.* The distinction is a most important one for the sweet gale, or bog myrtle, is a common and often abundant mire plant though usually of more mesotrophic mires than Bolton Fell Moss. The identification of *Corylus* is based on the lack of an enlarged endopore in the fossil material, a diagnostic feature of *Myrica* (Faegri & Iversen 1975, Godwin 1975) and other features of the grains (e.g. light staining rather than the deeper brown of *Myrica*). Many authors have recorded 'coryloid' or *Corylus* + *Myrica* curves where there is an admixture of the two types, but the author's case is based not only on grains of all the same type but also on the absence of *Myrica* from Bolton Fell Moss today (and probably in the past – Hodgson 1898), and from a number of 10 km^2 in the area (Perring & Walters 1962), and on the behaviour of the *Corylus* pollen curve.

Pollen of a number of families, genera and species were recorded which have not been included in the diagrams, mainly for reasons of clarity. For example, grains of *Centaurea cyanus,* once a common cornfield weed but now rare (Clapham, Tutin & Warburg 1962) were encountered, as were grains of the Caryophyllaceae, Umbelliferae and Papilionaceae; all such 'agricultural indicators' are aggregated into the 'Varia' curve as they are not strictly relevant, as separate entities, to this study. Similarly specific identifications were made within the genus *Plantago,* but aggregated for clarity in the diagrams.

The pollen of the Gramineae pose a special problem in pollen analysis, and in particular those species which could have grown on the bog such as *Molinia caerulea* and *Phragmites communis*. The special key of Faegri & Iversen (1964) is of some help in detecting possible *Phragmites* pollen (amongst other characteristics it is less than 26 microns in diameter) and using these characteristics it seems that this pollen type is rare. *Molinia,* however, is in a group of at least 13 genera and so cannot be distinguished on pollen analytical grounds. However, its clusters of tuberized stem-bases and sharply-twisted cord-roots were not found in the macrofossil analyses and, as already noted, it does not occur on the undisturbed bog at present. These observations, together with the way the Gramineae curve fluctuates in sympathy with the herb curves and inversely with the tree pollen curves, lead me to conclude that the grass pollen percentages are generally representative of clearance around the bog and not due to fluctuations of local grass populations. Such arguments must naturally be used with caution but the weight of palynological evidence in general is in favour of them.

Pollen and spores of most bog plants are readily identifiable and types such as *Sphagnum, Calluna,* other Ericales and Cyperaceae were all recorded, often abundantly (for example, 1,700 % *Sphagnum*). The key to *Sphagnum* spores of Tallis (1962) was not used, mainly because such observations were thought unnecessary in view of the extensive macrofossil work but also because the key is based on preparations using KOH only and mounting in glycerine jelly – and therefore includes divisions made on measurements of swollen spores. The key also does not differentiate between *S. cuspidatum* and *S. subsecundum,* nor between *S. imbricatum* and *S. magellanicum.*

All pollen counting was performed on microscopes of high quality – Zeiss, Leitz and Nikon – fitted with objectives of high numerical aperture. Routine traverses across 18 x 18 mm coverslips were done at magnifications of x320 to x400, critical identifications being made with oil-immersion objectives (including phase-contrast) at magnifications of x1000 to x1250. The non-random distribution of different pollen types on slides, demonstrated by Brookes & Thomas (1967), was borne in mind, and when it was not necessary to count whole slides, traverses were spaced to include representative areas of the slide.

4.3.4 Pollen sum

The general principles underlying the choice of the basis of the percentage calculations in a pollen diagram – the 'pollen sum' – have been reviewed in the literature (Wright & Patten 1963, Faegri & Iversen 1975, Barber 1976) but it will be appropriate here to consider the particular pollen sum used in this investigation. Until recently most British diagrams used a pollen sum of 150 AP (arboreal pollen); this has tended to change recently (Birks 1970, Hibbert *et al* 1971) and from the earliest diagrams (1967) the author has used a sum of 250 NMP – non-mire pollen. Had a sum of 150 AP been used it would have given an erroneous impression of high tree pollen percentages in what is a progressively deforested landscape, with one having to rely on artificially high Gramineae and other NAP percentages to correct this impression. These basic points are very well brought out by Faegri & Iversen (1975).

Working on the principle of excluding from the pollen sum local types liable to distort the diagram by over-representation, the non-mire pollen sum was arrived at. Spores of lower plants (e.g. *Pteridum aquilinum*), were also excluded in accordance with palynological convention. The pollen sum is therefore composed of all trees and shrubs, grasses and all non-mire herbs. Exclusions encompass all Ericaceae, Cyperaceae, all aquatics (e.g. *Menyanthes, Potamogeton* – capable of growth on the bog), all Filicales and Sphagna. It is recognised that some species included in the sum are capable of growth on the bog, such as *Betula, Pinus,* some grasses and some members of the Rosaceae (e.g. *Potentilla erecta*). However, as explained earlier tree growth is very limited (section 3.1.3) and birch is a common component of the forests of Cumberland (Godwin, Walker & Willis 1957), whereas pine was insignificant as a component until 19th century dry land plantations. Grasses have already been dealt with (sections 3.1.3 and 4.3.3) and families such as the Rosaceae contributed insignificant amounts of pollen to the diagram, and then a mixture of non-mire and possibly some mire species. In the course of establishing this pollen sum a number of papers dealing with the ecology of various species were consulted to see whether or not such species could grow in a bog habitat. For example, Sagar & Harper (1964) demonstrate conclusively that *Plantago* species cannot survive in *Sphagnum* bog.

A non-mire pollen sum seems then to be a fair way of illustrating the more or less regional changes in vegetation of the area surrounding the bog (see also Rybnicek & Rybnickova 1971), as well as being suitable for the correlation of former surfaces between different peat profiles because of the exclusion of purely local elements. A sum of at least 250 NMP was used for greater statistical reliability; this sum was met at all times and exceeded in particularly pollen-rich samples. It gave a total sum of usually 500 or so grains, and occasionally much more.

4.3.5 Pollen diagram presentation and correlation

A standardised format for the pollen diagrams was decided upon early in the investigation to facilitate correlation. This format is orthodox in following the 'normal' order of, from left to right, trees (birch, pine, oak, alder, hazel, etc.), AP v. NAP, grasses and herbs. It is slightly unorthodox in being a summary diagram, wherein a number of small herb curves have been aggregated into a 'varia' curve, so that the whole diagram can be reproduced at a manageable size – the individual components of the 'varia' curve are not important for correlation purposes. *Pteridium* was added to the diagrams on the basis of Oldfield's use of it (1960, 1963) as a clearance indicator; although there are some indications of its response in this way in the Bolton Fell Moss diagrams, doubt has been shed on its usefulness by Tinsley & Smith (1974).

The scales used in the pollen diagrams have been varied for clarity. A general scale of 1 cm = 10 % was used for the major components – most trees and grasses – but this was halved for the AP v. NAP curve, and doubled for the lower percentages of rare trees and for all herb pollen types. The vertical scale chosen is, like the macrofossil diagrams, a 10:1 reduction wherein each pollen sample can be shown at its 'true' thickness of 1 cm (= 1 mm line). The diagram type – bar-graph rather than 'resolved' or blacked-in curves, or the continental 'interaction' type – was chosen to give a truer picture of the levels counted and the individual percentages of the spectra, as well as to aid correlation.

Correlation of the diagrams was done by simply laying the diagrams, in their original tracing-paper form, over each other and manipulating them to achieve the best fit. As the diagrams do show a number of striking changes, consistent over a number of counts, this method was remarkably successful as well as simple and rapid. A straightforward zonation system was added to the diagrams to facilitate discussion.

4.4 OTHER ANALYSES

A variety of other techniques of analysis of the peat which might have proved useful in this investigation were explored. Some have been used to a limited extent as supporting information whilst others have not proved successful. The range of those possible measures are discussed briefly below.

4.4.1 Rhizopod analysis

The tests of the Protozoan Rhizopods are quite well preserved in peat and frequently encountered in pollen analysis if the preparation technique excludes hydrofluoric acid digestion of silica. Birks (1965) alerted the author to the possibilities of using these microfossils, which Heal (1961, 1962 and 1964) and others (de Graaf 1956, Paulson 1952) have shown to live in different microhabitats in fens and bogs. Tolonen (1966, 1971) has shown them to be of some use in interpreting bog stratigraphy, and his 1966 paper was used as a basis for identification (see also Barber 1976, figure 2.12). As HF digestion was not used on the pollen samples, rhizopod tests were counted along with the pollen and their relative frequencies expressed as a percentage of the non-mire pollen sum, rather than using the time-consuming separate analysis method of Tolonen (1966).

The results were disappointing. They added very little to the macrofossil analyses and were not all consistent in highlighting wet or dry phases in the bog's growth. Over a longer time span encompassing much more variable stratigraphy, including fen type peats, they may add some useful data and, significantly, this is the way in which Tolonen (1966) and Casparie (1972) have mainly made use of them. In the Bolton Fell Moss material only two species (*Amphitrema flavum* and *Assulina muscorum*) were at all common and therefore reliable as indicators, and so only one monolith's results were graphed (peat section HI, monolith 9). Possibly in future work on Bolton Fell Moss they may be useful in indicating micro-habitats within the deeper peat where macro-fossil Sphagna may not be so well preserved, but their inclusion in the present work was felt to be somewhat superfluous.

4.4.2 Iodine analyses

Through discussion with F.J.H. Mackereth the author became interested in the possibility of using iodine analysis as an index of oceanicity and rainfall variations. The basic idea is that iodine, together with other halogens (chlorine and bromine) are thrown into the atmosphere by bursting bubbles of sea water (Duce *et al* 1963, 1965), and, in the case of iodine, by direct evaporation (Miyake & Tsunogai 1963); they are then brought down to earth by precipitation and incorporated in humus and in plants. A greater proportion of westerly-type air masses, such as pertained during the Atlantic and Sub-Atlantic periods of the Flandrian, should then bring in more iodine to north-west England than would be precipitated in drier, more continental periods such as the Boreal and Sub-Boreal. To find out whether this sort of relationship was evidenced in the peat formed during the Medieval Optimum and the subsequent Little Ice Age iodine analyses were kindly performed by Jean Lishman of the Freshwater Biological Association, under the supervision of F.J.H. Mackereth, on two monoliths from Bolton Fell Moss.

Contiguous 1 cm slices of peat were analysed for iodine and total halide using an adaption of the method of Lein & Schwarz (1951) which is 'based on the catalytic effects of iodine ions on the oxidation of arsenious acid by ceric ions' (Pennington & Lishman 1971 – a paper which includes a very full consideration of the principle of the analyses with regard to lake sediments). In view of the possibility of different *Sphagnum* species accumulating iodine at different rates, analyses were also performed on fresh Sphagna collected from Bolton Fell Moss. It was found that samples of *Sphagnum cuspidatum* contained three times as much iodine as *S. rubellum* (7.54 micrograms per gram dry weight against 2.53 – mean values). A further problem was the possibility of iodine 'cycling' through *Eriophorum angustifolium* (Pennington & Lishman 1971) and the proven differences in iodine uptake between different species of vascular plants (Shacklette & Cuthbert 1967). Faced with these problems and not having available facilities for further analyses, only two profiles were completed (Results, section 5.5.3). The method needs a great deal more effort to develop it as a reliable technique for peat samples but it is hoped to follow this up in future work; the problems of dealing with much more homogeneous lake-muds are only now being fully understood (Pennington & Lishman 1971).

4.4.3 Silica

Angular fragments of silica were encountered in a number of pollen slides. Opal phytoliths (Dimbleby 1967, Rovner 1971) originating from plants were very few indeed and those found were plain and elliptical, and could not therefore be attributed to any particular group of plants such as grasses (Rovner 1971).

The mineral, rather than opaline, silica can only be windblown material brought onto the bog when soils are laid bare for cultivation or by natural means (Chapman 1964b). The former seems much more likely in view of the wholly Sub-Atlantic age of the profiles investigated, and in most pollen profiles such silica shows a peak abundance at the same time as maximum deforestation during the Napoleonic Wars period, c.1800 AD. Indeed at such levels the amount of silica was so great that difficulty was found in preparing satisfactory pollen slides, the silicone fluid residue being very gritty.

The amount of silica in each pollen preparation was estimated on the same 5-point scale used for macrofossils; an example of the results obtained is shown in macrofossil diagram HI9.

4.4.4 Fungal hyphae

Following Tolonen (1966) the author noted the presence of fungal hyphae in the pollen preparations. These presumably belong to *Omphalina sphagnicola* (Brightman & Nicholson 1966), which is common on Bolton Fell Moss today, characteristically inhabiting low mounds of *Sphagnum rubellum*. Tolonen (1966) notes that: 'Hyphae occur abundantly in the more decomposed layers . . . possible that the maximal occurrences denote drier phases in the development of the bog' (pp.156-7). This correlation is not at all clear at Bolton Fell Moss, high values (5-point scale) being found in various peat types, including hummocks, wet lawns and pool layers. Although recorded for all pollen counts this information was therefore thought unreliable and was not generally used.*

4.5 RADIOCARBON DATING

Absolute dating of major peat stratigraphical and pollen analytical events was seen as an essential part of this study, in that some previous studies have not

* Van Geel (1977) has since made a notable advance in this field.

had this advantage and have therefore been limited by a lack of an independent chronology.

Initially there were financial constraints and only two samples from section C could be sent for dating at the laboratory of Gakushuin University, Tokyo, but in 1969, as a result of their interest in the work, the C^{14} Laboratory of the Niedersächsisches Landesamt für Bodenforschung, Hanover, kindly agreed to date nine samples free of charge.

The principle and the laboratory technique of radiocarbon dating, and the recent 'calibrations' that have been applied to results, are not issues for discussion here (see West 1968, Barber 1976, etc.); attention is confined to the choice of samples and the mode of sampling.

As the peat strata were to be correlated using synchronous events in the pollen diagrams, five radiocarbon samples were taken from monolith HI9, the profile used for the master pollen diagram and one which was reasonably expected to have a record of continuous accumulation. Four further samples were taken from section HI; one on a lower pool layer more or less synchronous with a pool layer in monolith 9, and three from a monolith which passed from a strongly-growing hummock (one sample at hummock base) to a dried-out humified surface thought to be Medieval on pollen evidence (one sample) which was then flooded by a pool forming on top in response, it was thought, to the post-Medieval climatic deterioration (a further sample, which should have been synchronous with a pool dating from monolith 9). These nine samples from peat section HI were seen as the 'chronological key' to the work and it was felt sensible to concentrate them on one section, rather than to take dispersed single samples from several sections, so as to resolve any discrepancies in the results more easily. In the event this proved to be important.

The two previous samples from section C were taken from a single monolith (CI2) in order to date the lowermost pool in the section – the samples were separated by only 1 cm vertical difference. Unfortunately not enough carbon was yielded by one of these samples; the other gave a reasonable date.

All samples were taken with scrupulous regard to cleanliness and the avoidance of contamination. No carbon-based material was allowed to come into contact with the peat (such as polythene, oil, grease or paper), and the samples were taken direct from the aluminium monolith tins using specially made steel plates, 10 x 15 cm. Two plates were used for each sample, the peat lifted out between them and placed on a clean glass sheet. A 1 cm slice of peat was then removed from the edges of each sample and the remaining block of peat wrapped in aluminium foil. Throughout the peat was not touched by hand.

Sample size was increased from 2 cm thickness to an average of 3.5 cm following the failure of one of the section C samples for lack of carbon, giving a sample weight of between 130-200 grams (natural wet state), based on advice from the Hanover laboratory. Samples of this thickness would take about 50-60 years to accumulate, well within one standard deviation of the counting error on the radiocarbon date (minimum 70 years, average 120 years), and not, therefore, introduce any significant further error – a point discussed by Tolonen & Ruuhijarvi (1976). The results of this radiocarbon dating are discussed in section 5.3.

4.6 LAND USE HISTORY

The dating of episodes in pollen diagrams by correlation with archaeological cultures is a well-established technique used by all pollen analysts. It has not been superseded by radiocarbon dating but rather strengthened by it, the two techniques working together to shed new light on problems such as the earliest Neolithic phase in Britain. On a more recent time-scale, and using closely counted pollen diagrams, it has been possible to pick up historical episodes of short duration which have had an impact on vegetation and to characterize particular phases of post-Bronze Age agriculture (Oldfield 1963, Turner 1965, 1970). Oldfield in particular, in his 1969 paper entitled 'Pollen analysis and the history of land-use', has shown how episodes such as the Norse colonisation of the south-east Lake District (9th-10th centuries AD), the dissolution of Furness Abbey (1537 AD) and the subsequent rise of yeoman farmers, and the introduction of the New Husbandry (circa 1800 AD), may be recognised in pollen diagrams close-counted up to the present surface. Mitchell (1965) has based his post 3000 BC zonation of Irish pollen diagrams on such correlations with the historical record.

Furthermore the demise of particular species and the rise of others may be linked with land-use activities. Thus the general disappearance of *Artemisia* pollen is used to give a date of 1810 AD by Oldfield (1969) when deep ploughing was introduced into the south-east Lake District, and similarly the disappearance of *Cannabis* pollen marks the general importation of hemp from British India (Oldfield 1969, Godwin 1967). Re-afforestation during the 19th and 20th centuries also shows up as a notable feature in the uppermost centimetres of a number of pollen diagrams.

The recorded land use history of northern Cumbria was therefore studied in some detail, and that of the area surrounding Bolton Fell Moss in greater detail. The sources of this information are both published and unpublished; the latter were studied in the County Record Office, Carlisle, under the guidance of the county archivist. The general use and range of such sources used in this study are considered together with the related pollen diagrams in section 5.3, but one may make the general point here of the 'resolution' of land-use episodes in pollen diagrams. Such 'resolution' will be dependent upon the scale and duration of the land use event on the one hand, and dependent upon the rate of accumulation of the pollen-bearing sediment and the counting interval on the other. This problem has been explored by Turner (1970, pp.109

115) using a three-dimensional portrayal of selected pollen percentages from three diagrams from different parts of Bloak Moss, Ayrshire.

The study of land use history is therefore of relevance to this study, though one must beware of placing too great an emphasis on it – such is the intrinsic interest of this research that the point when the law of diminishing returns sets in is rather difficult to judge and one can find, as the author did, that much time is spent on gathering information which is beyond the resolution of the pollen diagrams.

4.7 CLIMATIC RECORDS

As previously noted, the possibility of linking independent evidence for climatic change with the peat stratigraphy was seen as one of the main advantages of work on Bolton Fell Moss. The primary sources of data on climatic change were the published works of Manley (for example, 1959, 1965) and of Lamb (e.g. 1966). Based on these, and guided by discussions with both authors, a card index of over 300 references was built up on the climatic changes of the British Isles and further afield, with particular emphasis on the period represented in the peat stratigraphy, the last two millenia.

In addition some local evidence was used. To test whether or not the Cumberland Lowland, part of which is in a well-known 'rain-shadow', could be regarded as an otherwise normal area exhibiting the same rainfall fluctuations as the England and Wales generalized record, a statistical analysis of local and national rainfall records was undertaken. The national record was obtained as an unpublished computer print-out from the Meteorological Office, by courtesy of H.H. Lamb; it is essentially an updating of the published record of Nicholas & Glasspoole (1931), bringing the record from 1727 to 1970. The local record stretches back only to 1861, and is compiled from records kept at Carlisle and Scaleby, 13 km and 6 km to the south-west respectively, and published in British Rainfall. The test for 'normalcy' involved isolating runs of 'wet' and 'dry' years from both records and testing the correlation between the two sets of records for the period 1864-1964. This established that the 1727-1970 record could be confidently used to characterize the broad rainfall fluctuations at Bolton Fell Moss, and, by implication, that periods beyond the instrumental record, known to be wetter or drier than average from a variety of 'proxy data' from elsewhere in Britain, could be assumed to be paralleled at Bolton Fell Moss.

Some interesting and unusual local data was also supplied by Cooper, Higgins & Partners, Consulting Structural Engineers, who were called in to advise on the settlement of the Norman tower of Carlisle Cathedral. They concluded that such settlement had occurred between 1250 and 1300 AD and was due to a sharp increase in the number of droughts.

In building-up and assessing the climatic record the author kept in mind the reliability and relevance of such records, and as with the land use history data, the resolution of climatic episodes in the stratigraphy. With regard to reliability one is, of course, dependent upon not only the modern authority but also on the original authority, especially for records beyond the time of instrumental recording (1670's for temperature, 1720's for rainfall). In general only the foremost authorities have been relied upon and the evidence cross-checked by referring to one or more authors. Relevance and resolution are intertwined. By the former is meant the relevance of, for example, the summer half-year record for rainfall and temperature which is presumably more meaningful for plant growth and peat accumulation than the incidence of extremely cold winters. This is not clear-cut, however – a cold, snowy winter would raise the water table, and hence the anaerobic 'sulphide' layer so important for active peat accumulation, more than a mild, dry winter. Similarly the odd extreme season, such as a single summer drought or a single very cold winter, may not be expected to be 'resolved' in the peat stratigraphy when the peat is accumulating at about 0.6 mm per year. The relationship is complicated by threshold factors leading to an apparently catastrophic event – for example, the relatively sudden appearance of pools over wide areas of the bog which could reflect the effect of a number of cool, wet years allied to a relatively impermeable humified surface. Such factors are discussed in later sections.

5. RESULTS

Two basic schemes for the presentation of results were considered: firstly, on an areal basis, reporting all the results from section A, section B, etc., and secondly, by way of the category of result – pollen, macrofossils, etc. The second scheme was chosen as seeming more logical, allowing easier comparison of the results of each stratigraphic section, and avoiding repetition. It also followed the rough order in which the research work had been done. After a section describing the location and general stratigraphy of each area of the bog investigated, the recorded field sections are discussed. The results of the pollen analyses used to correlate the stratigraphy, and radiocarbon dates, are then dealt with so as to give a chronological framework to the macrofossil analyses which follow. Finally the results of the study of climatic change records are given and correlations with the peat stratigraphy are suggested.

5.1 LOCATION OF PEAT SECTIONS

5.1.1 Area A

This area is located in the south-east quadrant of the bog just to the south of a main drainage ditch running from the central copse to the eastern edge of the bog (Map, figure 11). As noted in section 4.1.1, this peat section, and, in fact, the area itself was chosen more or less at random during the first visit to the bog in 1966. At that time areas A and B were the new active workings of the peat-litter company (A first cut in 1961, B in 1966) and seemed to offer the best opportunity of studying fresh peat faces which could be related to the modern surface; the surface between cutting ditches still displayed remnants of the original vegetation cover including Sphagna which had disappeared in that part of the moss cut since 1958 (the area containing the later-designated sections C, D, G and H).

Area A was defined as an area of peat cuttings covering roughly 100 x 50 m (this average size of 'area' was adhered to throughout) and all the peat faces in this block of land were inspected and the general stratigraphy noted and sketched (plate VIII). The average depth of cutting was 30-50 cm and over the most part it could be divided into an upper, relatively unhumified *Sphagnum* peat (generally H3) of 30-40 cm thickness, above a more humified peat of *Sphagnum-Eriophorum* and *Calluna* (H6-7). Broad shallow pools alternated with some equally broad, low hummocks, punctuated by steep-sided small hummocks of *Eriophorum* with some *Calluna.* The significance of a thin band of humified peat near the surface was not appreciated at this time. The peat face recorded, AII, is representative of this general stratigraphy; face AI, from which one of the first two monoliths taken from the bog had been extracted for the preliminary pollen and macrofossil work, had been cut back by the time stratigraphic recording took place, but the stratigraphy of the faces was more or less identical, there being only 25-30 cm between them.

5.1.2 Area B

Situated some 500 m to the south of area A, but still some 200 m from the present edge of the moss, area B is also one of relatively shallow peat cuttings. By the time this area was investigated closely (August 1967), it was known from pollen-analytical and documentary study that area A had been affected by the digging of the drainage ditch from the central copse to the eastern edge of the bog; hence the rehumified band of peat near the surface. There is no such humified band in the upper peat of area B and pollen analysis proves that the peat in section BI continued growth throughout the period of artificial retardation in area A (section 5.3). The uppermost peat of a large area of peat blocks at B is of fresh, unhumified *Sphagnum,* below which there are broad swathes of peat of various humifications, not dissimilar to those in area A – this again based on observations of an area measuring some 100 x 50 m. Isolated hummocks dominated by *Eriophorum vaginatum* also occur and a peculiarity of the general stratigraphy of area B is the large amount of *Sphagnum cuspidatum* occurring as broad swathes of greasy peat near the surface, showing that the whole area was recently very wet. This greasy peat gives rise to a characteristic horizon of 'pock-marks' as it dries out in summer; these are shown in the photograph of the general area (plate VIII), and the peat-face recorded is located by the marker poles in the upper right of the photograph. Unfortunately, as in area A, the peat face was cut back between the collection of the monolith (BI1) and the recording of the stratigraphy (section BII). This cutback, of only some 25-30 cm, produced only minor changes in the general stratigraphy (see section 5.2.2).

5.1.3 Area C

At the same time that area B was being investigated the large sector of the bog just cut in 1958, in north-west quadrant between central copse and peat works (see map, figure 11) was surveyed. Although none of the surface vegetation survived here, at least in a good state, with *Sphagnum,* this was unimportant in view of the deeper ditches which traversed the area, allowing a much greater time-span to be investigated and showing a number of interesting stratigraphic features. As it turned out all the sections worked on thereafter were from this part of the bog, the first in area C.

Ditches in area C averaged 70-90 cm, down to almost 2 m in the main north-south drainage ditches, which incidentally showed a very well-marked Recurrence Surface taken to be the 'Grenzhorizont' (RYIII) marking the opening of the Sub-Atlantic period at around 500 BC. It was, however, quite impossible to work in these ditches due to their depth and narrowness (circa 1 m), their generally high water level and the extremely soft bottoms. At this time these two parallel ditches were taken to be the surviving main drains of the 1800 AD reclamation attempt, shown on the first and all subsequent Ordnance Survey 10,560 scale maps, simply cleared out by the peat company in 1957. No information was given to the contrary by the peat company and this assumption was a costly mistake on the author's part for in fact the more westerly of the original drains had been left with its accumulation of soft peat and broken down walls, and a new drain fashioned from the old cuttings a few metres to the east – at least this is the most likely explanation, none of the original 1957 work force remembered any of this work. The details are considered further in section 5.4.2 dealing with pollen analyses, by which means the mistake was largely discovered.

In 1967, however, this disturbance to the west of the present main drainage ditches was not appreciated and area C (again roughly 100 x 50 m) was demarcated and studied. Attention was focussed on a section designated CI where it seemed that a double hummock system had overgrown a pool. All the sections in area C showed at least two distinct pool muds, unlike areas A and B; the upper one was generally within fresh *Sphagnum* peat (H3), transgressing over humified peat in places, while the lower one was often within, or immediately above, humified peat rich in *Calluna.* The stratigraphic situation generally was more complex than in areas A and B, with fewer broad swathes of a single peat type. Section CI is representative of this, even allowing for the disturbed nature of the uppermost horizons.

5.1.4. Area D

This area was originally defined as a block of cuttings over 200 m south of the northern edge of the bog. However, the peat in this area, on close examination, exhibited a rather odd granular structure with some signs of rehumification. Some attempt at reclamation for cultivation of this marginal area was therefore suspected and it was felt safer to abandon this area and redefine it farther south. It was thought advantageous to site area D as near as possible to the interesting area described by Ratcliffe in 1957 (section 3.1) and the SSSI notified by the Nature Conservancy in 1961 – this latter covers the whole of the south-west quadrant but the particularly fine part of it was around the angle formed by the meeting of the drains south from the peat works and west from the central copse (map, figure 11). The photograph (plate IX) shows the general situation of the area with the cut sections to the north and the SSSI beginning to the south; it was taken along a new ditch dug at a slight angle to the old 1800 AD drain marking the boundary between the north-west and south-west quadrants, as can be seen on the 1974 aerial photograph (plate III). The stratigraphy of area D in general shows a well-marked division into a lower humified peat (H6 and above), overlain by peat of humification 3-4 with a sharp junction between and with hummocks extending from the lower into the upper layer to varying degrees. The section recorded is representative of all the features noted in the general survey.

5.1.5 Area E

Although the north-east quadrant of the bog was known to have been drained and cut around 1800 AD on documentary evidence (see also 1951 aerial photographs, plate II, taken six years before modern cutting began) it was thought unwise to ignore this quadrant altogether. An area which, on the 1951 aerial photographs, seemed to have been less intensively disturbed than the rest was therefore surveyed as area E. Many of the ditches in this area were overgrown (plate X), making observations difficult in parts, but it quickly became obvious from both cuttings and ditches that the 1800 cutting, followed by intensive cutting since 1957, precluded reliable stratigraphic recording and that the uppermost peat was missing. Old infilled ditches were found, characteristically U-shaped with a succession of *S. cuspidatum, S. magellanicum* and *papillosum* and *Sphagna acutifolia,* but the old cuttings were not always obvious. A fair amount of fresh *Sphagnum* peat remained in the profiles, showing that the area had not been completely cut over to any depth, but due to drainage the topmost centimetres of peat were mouldered and oxidised and no peat beyond about 1800 AD had been formed except for the ditch infills, a fact confirmed by 'spot' pollen analyses. No further work was therefore done on area E.

5.1.6 Area F

This area, where the greatest peat depth of all was found (almost 11 m), is in the uncut south-west quadrant some 300 m due south of area D. The surface vegetation is of gently undulating *Sphagnum* lawn with a very high water table capped with *Eriophorum* and Ericaceae as described in section 3.1.3, and it was hoped to reconstruct the sub-surface stratigraphy in this area using short cores, closely spaced in a grid. This was before the failure of the same technique at Butterburn Flow and despite numerous attempts the technique failed to produce reliable correlations at area F also, and for the same reasons as given in section 3.2.1. Again some pool mud bands could be linked up and it was mainly due to the inability to securely correlate other types of peat that the idea was abandoned in favour of the much more fruitful studies of peat faces. It may be noted, however, that the pool muds encountered occurred at similar depths to those found in the peat sections and that a marked pool mud occurred over highly humified peat at around 180 cm in a number of cores both in area F and elsewhere, correlating with the depths of the probable main recurrence surface (RYIII; the 'Grenz') as seen in a few of the very deepest drainage ditches.

5.1.7 Area G

This was a much smaller area than those previously surveyed but was notable for its very well marked pool-muds and laminated *Sphagnum cuspidatum* peat. Area G is just to the north of area C (100 m north of section C) and the two faces recorded, GI and GII, were the first two faces to the west of the main drain (plate X). Examination of the 1951 aerial photographs revealed the possibility of this area being influenced by water tracks so that despite the probability of disturbance in peat so near the main drains it was felt worth recording the peculiar features of the sections, which included complex hummock-type peat as well as the pool types.

5.1.8 Area H

Following the realization, in November 1968, that the upper 30 cm or so of section CI was a consolidated mass of upturned turves from the 1800 AD drainage ditches, the author completely re-appraised the work and decided to search for a further section, hopefully unaffected by all drainage and peat cutting. This had to be in the part of the north-west quadrant containing areas C, D and G, for obvious reasons (mainly depth of ditches), and fortunately the peat company had cut back and cleared out a few of the drainage ditches running west to east across the sector. These had not been available for inspection earlier, being in an overgrown state similar to those in area E – overhung with *Calluna* 'bushes' with walls encrusted with algae and mosses and heavily weathered.

One particular drainage ditch, running for 370 m from uncut land on the west to area C in the east, showed a most interesting suite of stratigraphic features including hummocks of various morphologies as well as three fairly distinct levels of pool muds. Over most of its length this ditch was 90-100 cm deep – taking much of the drainage water from the uncut part of the north-west quadrant – and was paralleled by another ditch some 20 m to the north, the 'flat' between them being used for stacking the cut turves from the cuttings to the north and south. After a survey of both ditches a 7 m section of the south-facing ditch was selected as representative of the stratigraphy of this area H. The general features of the area are shown in the photograph (plate XI; a & b) in which the generally flat nature of the surface may be noted. One of the older peat-cutters could remember this area in an uncut state from the 1930's, when he used to shoot over it with his father. He gave a graphic, if unscientific, description of a typical regeneration complex with 'little pools, a foot or two across, and small rounded tussocks on which the birds used to sit'. He further remembered area G as being a 'bit of a quagmire' at this time, and the north-east quadrant as fairly dry with old, tall heather. Preliminary pollen analyses at 5 and 10 cm depth showed no evidence of overturned peat and so section H was made the main focus of further research. As well as the 7 m section, designated HI, a block of peat was investigated three-dimensionally using serial sections cut behind part of HI. These peat faces were designated HII to HVI.

5.2 FIELD STRATIGRAPHY

5.2.1 Sections AI and AII

Two monoliths were taken from area A on the first visit to Bolton Fell Moss in late 1966. No stratigraphy was recorded at that time and unfortunately, by the date of the next visit, the peat faces were found to have been cut back by about 30 cm. However, one of the poles marking the position of monolith AI2 had been repositioned directly behind the point from which the monolith had been taken (a prominent hummock still apparent in the cut-back section) and the peat face revealed by cutting (AII) was more or less identical to the previous face, on the basis of a field sketch made on the first visit. The pole marking the position of the other monolith – from a large pool in the peat face opposite face AI – had been stolen so a further monolith of the large poo

---

Plate VIII. a) General view of area A.
Plate VIII. b) General view of Area B, also showing 'pock-marks' in upper peat.
Plate IX. a) General view of Area C, looking north-east.
Plate IX. b) Area D, looking east to central copse and with uncut area to south. Note fall in level from uncut area to north.

Plate X. a) Area E, looking east. Note overgrown state of drainage ditch.

Plate X. b) Area G, to left of main drainage ditch.

Plate XI. Area H. The section recorded at HI is marked by the poles to the left, central copse is on the right. View conveys something of the size and surface regularity of the b

Plate XII. a) Stratigraphic section AII marked by poles.

Plate XII. b) Stratigraphic section BI marked by poles.

Plate XIII. Stratigraphic section CI. The position of monolith CI1 is shown cleaned, prior to sampling; the pool-hummock complex sampled by monoliths 2 and 3 is shown at the right.

Plate XIV. a) Stratigraphic section CII showing small cavities revealed by cut-back. The pool-hummock complex of section CI is still obvious to the right of the pole.

Plate XIV. b) Close-up of peat-face CII; the inverted layer is just above the knife.

Plate XV. a) Part of stratigraphic section DI showing top of hummock overgrown by fresh *Sphagnum* peat.

Plate XV. b) Stratigraphic section GII; corner of section showing inverted layer over fresh *Sphagnum* peat.

Plate XVI. a) Stratigraphic section HI: left-hand side of section showing persistent hummock and all three pool layers. Note that there is a shadow across the section.

Plate XVI. b) Section HI: close-up of monolith 9. Scale is 30 cm.

Plate XVII. a) Section HII. Middle pool-mud is clearly visible on left.

Plate XVII. b) Section HIV. Middle pool still discernible.

Plate XVIII. a) Section HV. Note central transgressive hummock.

Plate XVIII. b) West wall of pit (HI-HVI) showing middle pool mud and lower humified area.

to the left of the section was taken – this is monolith AII1. Both of these pool monoliths have very similar stratigraphy, the slight differences between them being mainly attributed to their not being from the same peat face even though they could be seen in the field to be from closely adjacent pools.

The stratigraphy of section AII (figure 14) begins with a widespread stratum of humified peat, generally about H6, rich in *Eriophorum* remains which show particular concentrations at 110 cm and 510 cm HS (horizontal scale). Over this humified surface an open water pool developed (410-625 cm HS), represented by a distinctive green-yellow greasy mud, which infilled with *Sphagnum* peat dominated by *S. cuspidatum* and some *S. subsecundum* (determined in the field with a x20 hand-lens). This wet mat of peat also overgrew the slight rise of humified peat on the far right of the section (625-690 cm HS) which, to judge from the lack of a pool mud, had stood slightly above the initial open water pool. A feature of this pool infill, not thought important at the time of recording but noted nonetheless, was the scatter of *Eriophorum vaginatum* tufts (identified by their blackened spindles in the leaf-bases) which became incorporated into the peat without forming noticeable tussocks. This observation was subsequently found to match the surface observations of Osvald (1923) who comments that the species may frequently be found as an early immigrant in such situations. It is also interesting to note that Walker & Walker (1961) record, without comments, the presence of the species in a similar situation in one of their sections (Clara bog, figure 7, page 181). The role of *Eriophorum* will be considered in this context and others in the discussion (section 6.3).

This large pool was bounded to the north by an *Eriophorum-Calluna* hummock of a diffuse nature (at 400 cm HS, 35 cm VS – vertical scale), incorporated into what must have been an actively growing island of hummock plants. This broad, flat-topped hummock maintained itself without overgrowth from the neighbouring pool and indeed shows signs of extending out over the infilling pool (420 cm HS) – unless this feature is to be interpreted as an overhang or 'cornice', a feature of some *Eriophorum* and *Molinia* hummocks which is familiar to most bog ecologists. To the left of the section this broad humified surface terminated and graded into a hummock rich in *Calluna* and *Eriophorum* and of slightly higher humification (H7 against H6), beyond which there were two isolated *Eriophorum*-rich tussocks, one of which is overwhelmed by the unhumified peat growing up around it, the other surviving to become incorporated into the widespread humified surface at 5-10 cm VS.

It was noted that the prominent hummock at 110 cm HS had been partially overgrown by fresher *Sphagnum* peat, having presumably stagnated while the hollow to the left infilled, and to examine the details of this overgrowth process and correlate it with the infill of the large pool, a monolith was taken from this hummock; the results are reported in sections 5.3 and 5.4.

The dark band of humified peat, averaging H7, coloured dark-brown to

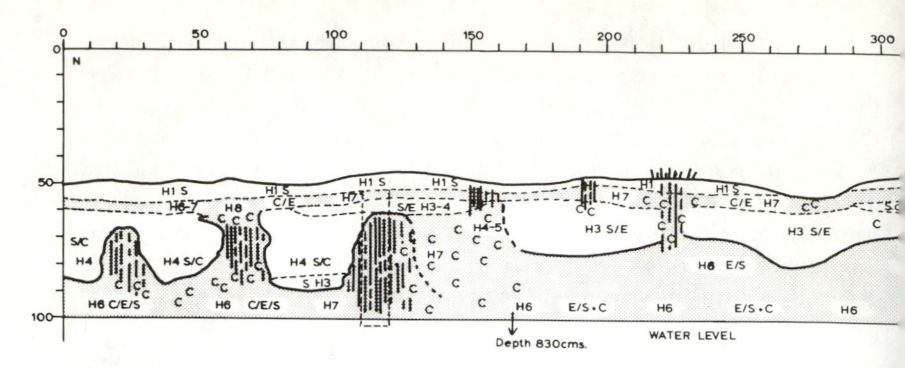

Figure 14. Field stratigraphy of section AII.

black and rich in *Calluna* twigs, was suspected as being artificial from an early stage in the investigation. As can be seen on the photograph (plate XII) it shrank on drying out in summer and transgressed hummocks, hollows and flats indiscriminately. Between 570 and 650 cm HS this band had a distinct black crumbly texture and charred *Calluna* twigs were evident, as if a small fire had affected the peat. As this section was near the parish boundary running west to east across the moss it was assumed that the digging of the boundary ditch had dried out the adjacent peat, perhaps also causing secondary humification of the surface peat. The date of this was estimated as early 19th century on the map evidence then available (this dating is considered further in later sections) and it was therefore looked upon as a potentially useful marker horizon. There are some slight signs of pool formation just below this horizon (*S. cuspidatum* lens at 300 cm HS), and above it a thick loose carpet of unhumified *Sphagnum* peat accumulated. Through this carpet a number of *Eriophorum vaginatum* tussocks seemed to be forming (at 220, 370, 550 and 650 cm HS) and both *Calluna* and *Erica tetralix* were common on the surface.

Section AII may therefore be summarized as being peat with no convincing evidence of cyclic alternation over the time period represented by the 50 cm of peat (estimated in 1967 to be circa 700 years) but showing instead growth of peat in broad swathes of wetter and drier communities, punctuated by occasional, largely *Eriophorum vaginatum*-based, hummocks, the major 'event recorded being the flooding of the relatively dry surface at the base of the profile and the eventual overgrowth of almost the whole section by *Sphagnum* peats of low humification. It is difficult to see how this can be accounted for, other than by a change in the overall hydrology of the bog.

5.2.2 Sections BI and BII

The stratigraphy of area B exhibited a four-fold division of broad swathes of

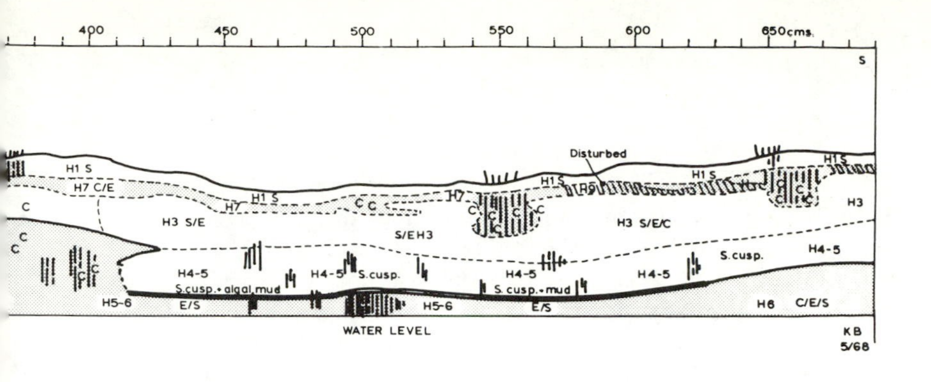

peat, beginning with a widespread drier type of community rich in *Eriophorum vaginatum* and *Calluna.* This stratum appeared as low mounds (see figure 15, section BII, 280-400 cm HS, 40 cm VS), at a number of points at the base of the cutting ditches and could be located a few centimetres below the ditch bottom in other places (e.g. at 250 cm HS on section BII). From this widespread drier surface a number of isolated steep-sided hummocks appear to have survived and maintained themselves within a matrix of low humification *Sphagnum* peat which grew over the dry surface. There was no sign of any lenticular or other structures within this broad band of H3 peat. At about the time when this *Sphagnum* lawn was growing level with the tops of the isolated hummocks, at about 20-25 cm below the surface, the whole area evidently became wetter and supported a community rich in *Sphagnum cuspidatum,* giving rise to the band of greasy, somewhat felted *S. cuspidatum* peat which dries out a characteristic light-coloured, pock-marked horizon, as seen in the photograph (plate VIII). This layer also contained much *Eriophorum,* both small caespitose growths of *E. vaginatum* and isolated rhizomes of *E. angustifolium.*

The *Calluna-Eriophorum* hummock at 50 cm HS underwent some rejuvenation at this time with *Calluna* spreading outwards and upwards, but was soon overwhelmed by the wet *Sphagnum* lawn. As previously noted the peat was cut back between the monolith being collected and the stratigraphy being recorded, but the monolith position was recorded by a pole which, being 30-40 cm behind face BI, was undisturbed by the cutting. Comparison of the stratigraphy of monolith BI(1) and section BII, and the stratigraphy of BI as sketched on the first visit, clearly showed the monolith to be from the side of the hummock cut through by section BII, and as the pollen analyses had already been done on monolith BI(1), it was also used for the macrofossil analyses to demonstrate the details of hummock overgrowth. Following this overgrowth another broad swathe of fresh *Sphagnum* peat was laid down, with a few knots of *Eriophorum vaginatum* at 150 and 180 cm HS; from still living

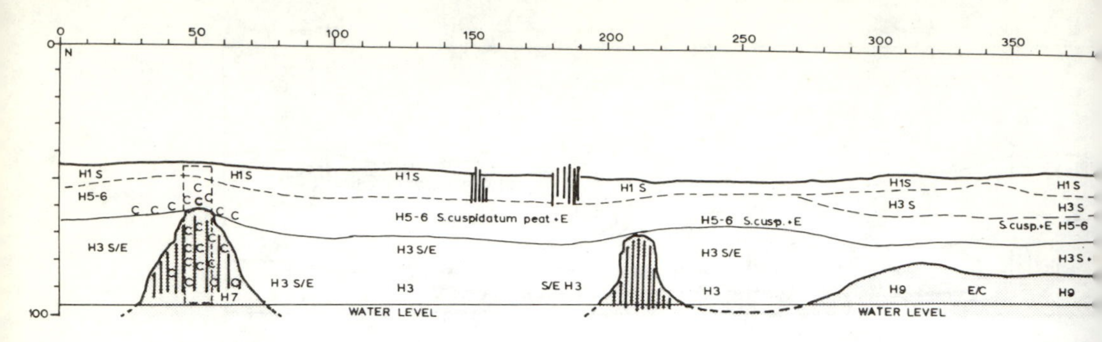

Figure 15. Field stratigraphy of section BII.

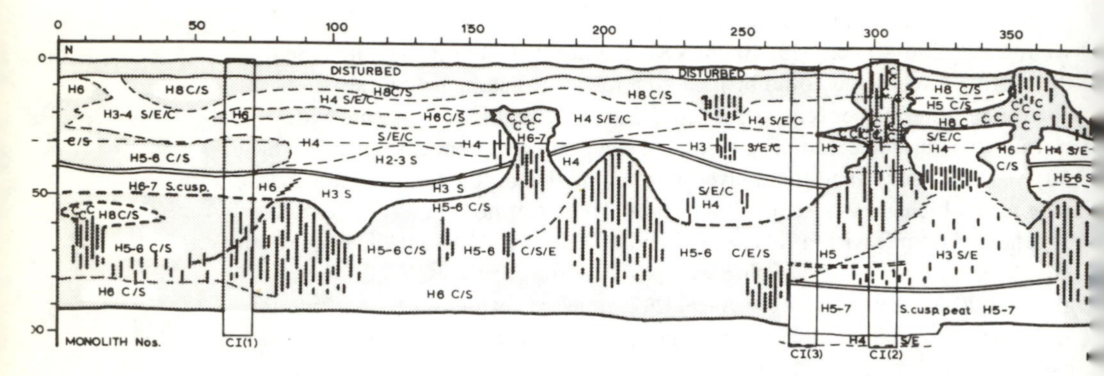

Figure 16. Field stratigraphy of section CI.

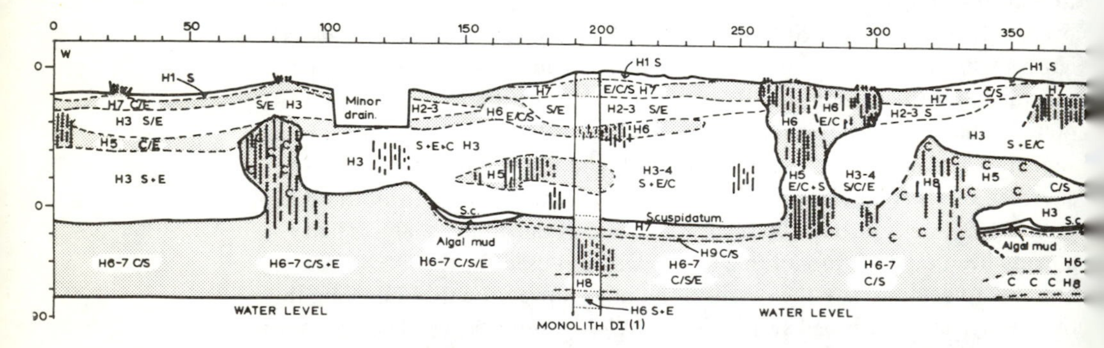

Figure 17. Field stratigraphy of section DI.

specimens on the present surface much of this H1 *Sphagnum* peat was seen to be made up of *Sphagnum magellanicum,* along with one or two species of the Acutifolia group.

As with section AII it is difficult to account for the above succession of form and type of peat by means other than an increase in wetness at the bog surface.

5.2.3 Sections CI and CII

As mentioned in section 5.1.3 the peat face designated CI was of interest in exhibiting the pool muds and an apparent example of hummock-hollow inversion as predicted by Osvald (1923). This phenomenon is not affected by the later discovery that the stratigraphy above about 30 cm is made up of overturned sods of peat.

Accumulation in section CI (figure 16) begins with a generally humified *Sphagnum-Calluna* complex at 90-80 cm VS, except for a *Sphagnum cuspidatum* hollow between 270 and 370 cm HS, bounded by two *Eriophorum* tussocks. While the area to the north (left in the diagram) grew up as a moderately humified and fairly uniform stratum, rich in *Calluna* remains (0-270 cm HS, 90-60 cm VS), the southerly hollow is converted into an open water pool, represented by a greasy yellow-green pool mud at between 80 and 85 cm VS. The infill of this pool by weakly humified *Sphagnum* peat was accompanied by numerous attempts at immigration by *Eriophorum vaginatum,* as noted in surface observations by Osvald (1923). Some 10 cm or so above the first pool mud a second less distinct one formed, again indicating open water and also terminated to the north by the same *Eriophorum* tussock. To the south, however, it terminates at 310 cm HS, 70 cm VS against no such obvious obstruction and we may assume that a small boss or mound of unhumified *Sphagnum* peat impeded its spread. At the same time this boss of *Sphagnum* infilling the pool was being 'encroached' upon by fast growing, moderately humified peat, rich in *Eriophorum,* as shown in the stratigraphic diagram CI (figure 16) and in the photograph (plate XIII). The whole section at this period appears to be in a hummocky mode of growth with centres of *Eriophorum-Calluna* hummocks at 10, 80, 160-200, 300 and 370 cm HS. An irregular more or less humified surface appears to have been formed at about 50 cm VS, probably representing a general slowing down in the rate of peat accumulation; certainly there was a complete absence of open water, the small hollows which may have supported damper *Sphagnum* communities (at 100 and 240 cm HS) showing no evidence of pool muds or greasy *Sphagnum cuspidatum* layers.

This irregular surface was then covered by a renewed growth of Sphagna, including *S. cuspidatum* over humified peat at 0-80 cm HS, culminating in a marked pool mud, obvious on the photograph (plate XIII), at around 30-40 cm VS. This pool terminated against the hummock at 300 cm, reappearing at

360 cm HS, 35 cm VS, and probably encircling the hummock complex which had grown up over the previous pool – the area between the two hummocks was occupied, as indicated on the stratigraphic diagram, by a dense growth of *Eriophorum vaginatum.* Both hummocks show indications of being almost swamped by this widespread pool: within the hummock peat at the level of the pools there are slight traces of greasiness in the peat and very small patches or flecks of fresher lighter-coloured Sphagna. However these were not sufficient to threaten the basic integrity of the hummock forms which, although they may only have existed a few centimetres above water level, continued to grow up as recognisable entities, rich in *Eriophorum* and *Calluna* – as is shown in the macrofossil analyses from monolith CI(2). So too did the small hummock at 170 cm HS, which may be seen as a break in the pool mud just to the right of the spade in the photograph (plate XIII). The wavy form of the pool mud may not, of course, accurately reflect the base of the actual pool because of subsequent disturbance by ditch digging and distortion from the dumping of sods of peat over it, but if it does accurately reflect the form then one may postulate either a deep pool with an undulating bottom, or a pool which grew and encroached over the surrounding peat, for example, the pool represented by the mud from 180 to 280 cm HS.

In any case before the pools could infill very far – with, in one case (80-150 cm HS) an interesting mound of fresh *Sphagnum* reminiscent of the lenticles of von Post & Sernander (1910) – a ditch was unfortunately dug alongside the section and the spoil thrown up onto the old surface, so that, from pollen-analytical evidence and the evidence of the stratigraphy of peat face CII, all the rather interesting stratigraphy above about 30 cm is artificial.

Three monoliths were taken from section CI. The first, CI(1), whose position may be seen as the cleaned area on the photograph (plate XIII) was taken to examine the succession from hummock to pool, while CI(2) was taken to show the infill of the lower pool and the growth of the hummock over it. It was later decided to take monolith CI(3) to investigate in more detail the relationships between pools and hummock between 270 and 310 cm HS. The details of the analyses on these monoliths are reported in sections 5.3 and 5.4.

The peat section designated CII was found to be exposed on a field visit in November 1968. Pollen analyses on the monolith from section CI had already given anomalous results, but confirmation of the artificiality of the upper horizons in area C was brought home to the author when examining section CII (photographs, plate XIV). The jumbled nature of the upper stratigraphy was more noticeable in section CII, being fresh and less weathered than CI, and just above the pool mud a number of small holes (2 x 0.5 cm and less) were noticed. On probing some of these turned out to be the entrances to small cavities behind the peat face, undoubtedly formed by the piling up of the cut sods when the old ditch was dug. The close-up photograph (plate XIV) shows the junction of undisturbed and overturned peat, just above the knife, and one of the small cavities exposed by the author. As can be seen from this photograph the distinction between the two types of peat was by no means clear, and even in the monoliths from section CI this horizon could only be positively identified by pollen analyses from each side of the border.

No stratigraphic record was made of section CII because of time limitations and the need to find an undisturbed deep section elsewhere, and because the cutting itself was shallow.

The truncated record of section CI is nevertheless of much interest with its superimposition of pools and hummocks, and details of the chronology and succession of species are reported and discussed in later sections.

5.2.4 Section DI

The section (figure 17) was recorded from a drainage rather than a cutting ditch, in order to examine a greater depth of peat. The lowermost 25 cm or so of peat is all moderately to well humified (up to H8-9 in places), relatively rich in *Calluna* twigs, though not in *Eriophorum,* and of an even, cheesy texture. This peat must represent some centuries of growth by relatively dry lawn communities and the thin H9 layer, below the *Sphagnum cuspidatum* regrowth, may represent an actual cessation of accumulation or minor retardation layer in the sense used by previous authors (Godwin 1954). The hummocks at 80 cm, 270 cm and 320 cm HS may be assumed to be in existence then, at least in embryonic form, and stood proud of the general surface when it was flooded and occupied by *S. cuspidatum* and some open water. This flooding event is widespread and at about the same depth (45-52 cm VS) throughout area D and probably equates with the major changes observed at around the same level all over the moss. Pool muds are not prominent in the section except at 140-170 cm HS and 340-400 cm HS; elsewhere they are very thin or discontinuous indicating the rapid infill of any open water. This infill and build-up of a wet *Sphagnum* lawn is, like the peat below the retardation layer, of an even-textured nature with no indications of lenticular structures; the hummock which appears 'suspended' in the section at 150-205 cm HS, 30-40 cm VS, was shown by excavation to be an outgrowth of a hummock behind the line of the section which was traceable back into the humified surface – that is, it had not spontaneously arisen within the infill peat.

The three main hummocks have rather different histories. That at 80 cm HS, a boss rich in *Eriophorum vaginatum,* appears to have maintained itself for some time and to have contributed to a rather more humified stratum (H5-6) at about 20-15 cm VS over the H3 hollow infill. It was then overgrown by a wet lawn community giving rise to a lightly humified carpet of Sphagna, in which some remains of *S. cuspidatum* could be recognized in the field. This phenomenon can be seen in the close-up photograph (plate XV) which illustrates the topmost stratigraphy of the section from 0-150 cm HS – no photograph of the lower stratigraphy could be obtained at section D because of the narrowness of the ditch. It will be noted that this hummock overgrowth was not the result of the infill of any discrete hollow beside the hummock but was caused by a general move to a widespread wet lawn formation which is apparent across most of the section DI and elsewhere in area D.

In contrast, the hummock at 270 cm HS maintains itself up to the present surface (right hand side of section, figure 17) and even expands as it nears the surface, growing over a hollow between it and the hummock centred at 320 cm HS. This latter hummock, rather more diffuse in nature and less dominated by *Eriophorum vaginatum,* but nevertheless a definite entity of H5-8 peat within H3 peat, was obviously also present at the time the bog surface supported the pool from 340-400 cm HS. It then appears to have colonized the infilling pool, (a 45 cm overhang being most unlikely) before being progressively overwhelmed by the damp *Sphagnum* lawn represented by the H3 peat.

The dark H7 peat band near the present surface could well be due to drainage attempts during the 19th century (section DI is some 70 m from an old infilled drain) and the H1 fresh green *Sphagnum* still remaining at a number of places represents regrowth after the choking up of the same drain.

A single monolith was extracted from the section at 190-200 cm HS; this point was chosen as it covered all the strata observed and as these strata were interrupted by the various hummocks (for example the humified layer from 0-240 cm HS, 15-20 cm VS, and its relationship to the hummock at 80 cm HS) pollen analyses on this monolith could be used to indirectly date the stages in hummock growth.

Once again it may be surmised from the evidence of the field stratigraphy alone that the mode of growth exhibited by section DI is in no way cyclical and that, over a period of several centuries, the communities present were of a widespread lawn type, through which hummocks persisted for greater or lesser periods of time.

5.2.5 Sections GI and GII

The stratigraphy of section GI (figure 18) is notable for its laminated *Sphagnum cuspidatum* peat, of humification 5-8, and as mentioned in section 5.1.7, it is thought that this peat accumulated in an area of the bog subject to some water movement. Borings below ditch level at 30, 50 and 70 cm HS revealed that the laminated peat was occupying something of a semi-circular pool or possibly channel. The peat on either side of this pool at this early stage in section GI was highly humified and formed a sharp boundary to the pool; it seems possible from the form seen in section that the pool or channel undercut these two humified banks. The central humified area between 70 and 280 cm HS was in any event flooded and a distinct pool mud was laid down. Above this three small hummock remnants protruded for varying times, the central one (165 cm HS, 65 cm VS) was soon submerged under laminated,

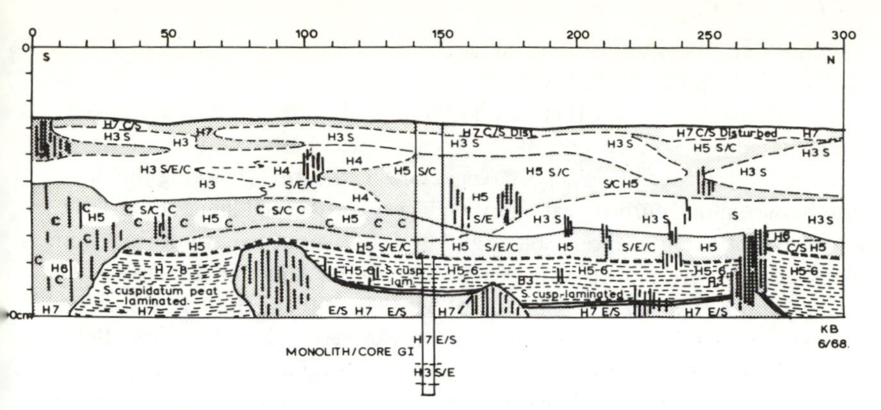

Figure 18. Field stratigraphy of section GI.

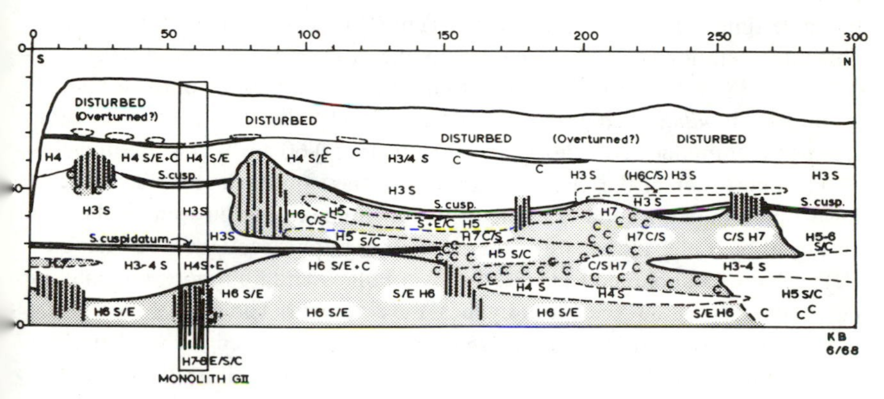

Figure 19. Field stratigraphy of section GII.

greasy *S. cuspidatum* peat, while the more southerly hummock (80 cm HS) survived almost until the pool was infilled. The more northerly hummock, in reality a tussock of *Eriophorum vaginatum,* survived to contribute to a more humified, drier surface of cymbifolian *Sphagnum-Eriophorum-Calluna* which grew over the infilled pool, spreading from the persistent hummock form between 0-30 cm HS. The stratigraphy above the sharp boundary at about 30 cm VS (0-120 cm HS) down to 40 cm VS (120-300 cm HS) is thought to be in overturned peat, at least in part. This conclusion is based on pollen-analytical evidence (section 5.3), on the nature of the peat in the monolith when closely examined in the laboratory – peat of very variable texture and bands of different humification – and on reconsideration of the field stratigraphy, for example the odd formation at 60-130 cm HS, 10-30 cm VS. The monolith and core (taken with a Russian peat sampler as no tins were available) were taken to investigate the timing and character of the flooding of the hummock at 64 cm VS, the infill stages of the pool and the spread of a drier community again at 50 cm VS.

In the case of section GII (figure 19) some 5 m to the west of section GI, the artificiality of the uppermost 15-20 cm of peat was obvious from the start, as the photograph (plate XV) makes clear, the knife marking the junction between undisturbed peat and the gingery overturned layer above. The lower part of this peat face displays an interesting sequence of pool and hummock peats, the earliest stratum being a more or less uniform dry lawn of H6 *Sphagnum-Eriophorum-Calluna,* except between 260 and 300 cm HS where it merges with a slightly less humified peat relatively richer in *Calluna.* Between 60 and 150 cm HS this lawn appears to have built up into a low hummock with damper *Sphagnum* hollows on either side. The H4 *Sphagnum* lens to the right of this hummock (160-260 cm HS, 65-70 cm VS) was succeeded by a drier community, rich in *Calluna* which laid down peat of humification 7 at the same time as hollow to the left (0-90 cm H5) was infilling with H3 peat. This interpretation is based on the datum line of the pool mud stretching from 0-150 cm HS, at 60 cm VS, across both hollow and hummock until stopped by humified *Calluna*-rich peat. The wedge of H3-4 *Sphagnum* peat at the other side of the hummock complex (225-300 cm HS, 50-60 cm VS) is presumably also an expression of this change to greater wetness on the bog surface. The lowest pool-mud is a good example of the phenomenon often seen by the author in the field, and recorded on other stratigraphic sections (for example HI) – that of a pool mud transgressing over hollow and hummock communities quite contrary to the concept of cyclical development. The central core of the hummock, of H5-7 *Sphagnum-Calluna* peat, which separated the pool and wet lawn just described, grew outwards and upwards to establish itself over parts

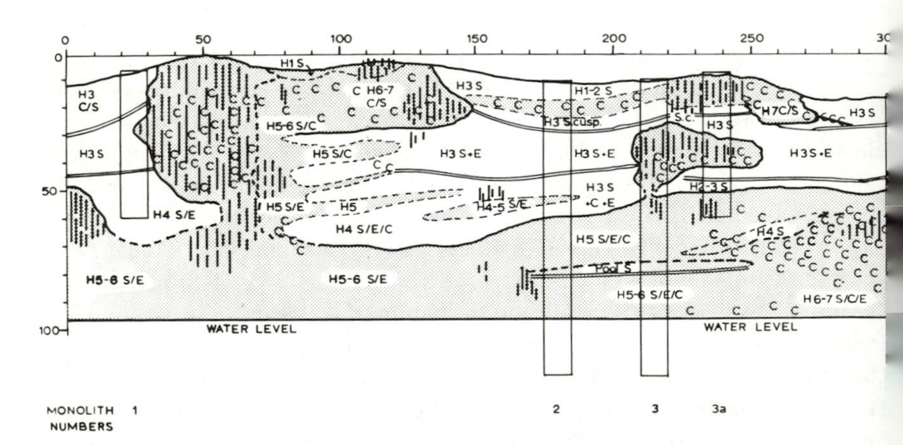

Figure 20. Field stratigraphy of section HI.

of both the wetter communities, so that at its maximum it stretched from 70 to 280 cm HS. The first 70 cm at the left of the section remained wet however, the pool infilling with fresh, little-humified *Sphagnum* peat, as can be seen on the photograph (plate XV).

Both hollow and hummocks grew up at approximately the same rate, as evidenced by the extensive and presumably more or less simultaneous development of pools at about 30-40 cm depth (these pools continue for tens of metres at the same level along the ditch). Although some hummock tops survived this flooding they soon became overgrown by the wet carpet of Sphagna which accumulated right across the section. The odd outgrowth of humified *Calluna*-rich peat which grew across part of this *Sphagnum* lawn (from 200-270 cm HS, 30 cm VS) is possibly related to a fire on the moss (see macrofossil analyses, section 5.4). Finally pools were again forming in section GII, above the *Sphagnum* lawn, between 10 and 80 cm HS and 150-200 cm HS, when they were sealed in by overturned peat in 1800 AD thus becoming a valuable marker horizon. The monolith from this section was taken at 60 cm HS so as to pass through all three pool muds.

5.2.6 *Section HI*

As already noted in section 5.1.8, this peat face was chosen for detailed study, after the disappointments with section CI, because of its depth, its lack of disturbance and the complexity of its stratigraphy. It showed all the features apparent in a survey of the whole ditch including a number of different hummock forms (figure 20). It was also decided at an early stage that the section

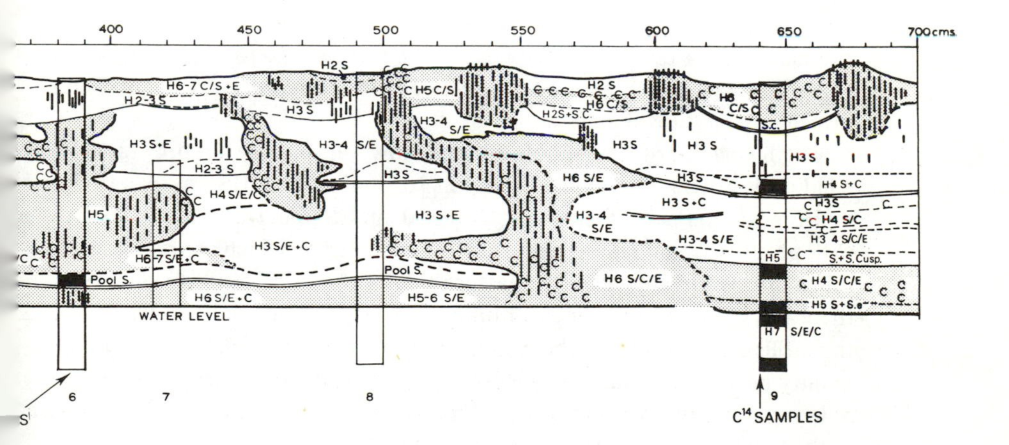

between 200 and 400 cm HS, containing an alternation of pools and hummocks, would repay investigation in three dimensions. Accordingly a number of serial sections were cut back in the peat face; their stratigraphy is reported in the next section, 5.2.7.

The earliest peat visible in section HI is once more a quite highly humified one, indicative of relatively dry conditions and dominated by cymbifolian Sphagna and varying but high proportions of *Calluna* and *Eriophorum vaginatum.* It may be noted that this type of community was also the commonest in the monolith stratigraphy below water level and that pool muds were not present in the submerged parts of monoliths 2-9, nor was there any pool mud or even any evidence of wet conditions in the core taken below monolith 9 in order to extend the pollen diagram to 150 cm. The next pool mud was found at a depth of 185 cm, and this gives about 100 cm of peat of H5-9, possibly representing a drier climatic period of some considerable duration.

The change from this broad surface of humified peat to less humified peat, a major change affecting most of the section, occurred first (that is at the greatest depth) on the extreme right of the section between 620 and 700 cm HS, 85 cm VS. A greenish-brown greasy band of peat, containing obvious *Sphagnum cuspidatum* but not a well-developed pool mud, lies above dark brown peat with much *Calluna,* and it is the precursor of a remarkable development of a persistent hollow or flat with no fewer than five pool phases. We may assume the existence of a slight mound of humified peat at 620 cm HS, the beginning of the surviving part of the dry surface covering the rest of the base of the visible section. The next stage must have involved the building up of low hummocks between 240 and 340 cm HS, 65-75 cm VS; at 550 cm HS, 75 cm VS and a broad raised area between 0 and 165 cm HS, 70 cm VS. Only then could the other two lower pools (165-240 cm HS and 340-550 cm HS) have formed. Both pools have only a poorly developed mud band but the peat throughout was very greasy, indicating shallow, *Sphagnum*-choked pools with abundant algal growth on the Sphagna – in the field *S. cuspidatum* was identified from both pools as was *Sphagnum cf subsecundum,* subsequently confirmed by macrofossil analysis.

The formation of these pools did not, however, presage a return to wetter conditions generally. The pool between 165 and 240 cm HS was overgrown by H5 peat, and part of the larger 340-550 cm HS pool was directly overgrown with peat of around H6. After a short interval during which some H3 cymbifolian *Sphagnum* grew over the eastern side of this pool (510-545 cm HS) there was some outspreading of hummock species, particularly *Calluna,* over the infilled pool, but over the central part of the pool a wet *Sphagnum* community persisted, giving rise to H3-4 peat of remarkable purity – that is a *Sphagnum* mixture with very little *Eriophorum* or *Calluna* detectable, at least in the field – as seen in the photograph, plate XVI (the macrofossil results bear this out – section 5.4). At about the same time hummock formation seems to have begun in earnest with centres at 5, 60, 210, 300, 380 and 550 cm HS and a broad dome of humified peat accumulated between about 150 and 420 cm HS. In the HI9 hollow (600-700 cm HS) peat accumulation stayed generally on the unhumified side after the major change at 85 cm, varying between damp lawn conditions and pools, with an occasional *Calluna* band, as between 48 and 53 cm VS – this absence of any hiatus and the good state of preservation of pollen and macrofossils, as well as the fact that it contained five pool layers, made monolith HI9 an obvious choice for the master pollen diagram, for five radio-carbon dates and detailed macrofossil analyses.

The next discernible stage in what was probably a continuous growth of peat is the widespread dry surface prior to the pool layer at around 30-40 cm VS. This dry surface is best developed between 210 and 420 cm HS (circa 40 cm VS) where it formed an almost level surface (photograph, plate XVI), probably contemporaneous with the others at this level. The 'outcrop' of H5 peat and the lens of H4-5 peat in the hollow between 90 and 200 cm HS seem reasonable candidates for the contemporary surface at the time of maximal dry surface extension further to the right. The persistent hummock centred at 60 cm HS would by this reasoning therefore have reached at least to 50 cm VS and possibly higher; the hollow to its right would be infilled to something slightly less than the same level.

The striking pool development of 30-40 cm VS cuts across the stratigraphy irrespective of hummocks and hollows. Over the central dry flat it is noticeable that the pool mud does not lie directly on the humified peat but is separated from it by a few centimetres of fresh H2-3 peat (plate XVI) a feature reminiscent of the 'precursor peat' of Recurrence Surfaces (Godwin 1954). The pool development at this depth represents the wettest phase in the recent history of the bog – of the 7 m section some 370 cm were open water on the evidence of the pool muds and the rest must have been very wet lawn or rather unstable hummock – for example the hummock at 210 cm HS had only just survived the flooding and that at 460 cm HS was without firm foundations. The synchroneity of the pools seems obvious from the field stratigraphy but was tested as were the lower pools by pollen-analytical correlation (section 5.3). The length of time that a pool remained open water and how long it takes to form a centimetre of pool mud are more difficult to assess. Observations of the state of preservation of both pollen and macrofossils from the pools, the thickness of the muds and considerations from elsewhere (for example, Boatman 1977) lead one to think that these pools were shortlived – that is, existing as open water for a decade or two rather than a century or more. Further consideration of this question is given in section 5.3.

Infill of these pools was exclusively by means of only slightly humified *Sphagnum* peat (H2-4) and it is interesting to note that at a number of points (430, 500 and 610 cm HS) the first infill was by a lens of almost pure *Sphagnum* peat (see macrofossil results of monolith HI8, section 5.4). This broad swathe of *Sphagnum* peat grew up around the remaining hummocks which appear to have colonized part of the lawn as conditions became more suitable for hummock species such as *Calluna* (for example, hummock extensions at 200-250 cm HS and 370-330 cm HS). On the whole though the *Sphagnum* carpet was very wet, to judge from the freshness of the peat and the presence of *Sphagnum cuspidatum* (determined in the field with hand lens; confirmed in macrofossil analyses – see monolith HI4, section 5.4). The possible correlation of these hummock extensions with known climatic changes is dealt with later.

The fate of individual hummocks varied; that at 50 cm HS persisted up to the present surface and became the nucleus for a dry *Calluna-Eriophorum* community which spread out to 150 cm HS, obviously coincidentally with the formation of various other hummocks at 540, 600 and 680 cm HS, in that they are all related by the upper pool layer. The hummock covered by monoliths 3 and 3a (200-250 cm HS) was clearly overgrown by the *Sphagnum* lawn before the pool above it formed and this interesting phenomenon was investigated by close macrofossil sampling. The hummock sampled by monolith 6 is interesting in that it appears to have been swamped by a wet lawn or pool while still actively growing itself – that is to say there are no obvious signs of hummock degeneration or cessation of growth prior to its overgrowth by a different community. The hummocks at about 450 and 500 cm HS, although almost overwhelmed, managed to grow through the wet *Sphagnum* carpet and contribute to a drier plant community near the present surface. This drier community, rich in *Calluna* is in fact part of the final phase of peat growth in section HI, which may be taken to begin with the formation of the upper pools at depths between 10 and 18 cm below the present surface.

Not all of the upper pools have a mud base, and in those which do, the pool mud is not so strongly developed as in the middle pool layer, indicating that the upper pools were probably shallower and/or shorter-lived. Above the mud, or as a first stage in those pools without mud (for example, between 350 and 400 cm HS, or 550-600 cm HS), there is a fresh green layer of *Sphagnum cuspidatum,* usually with some other species present, as the macrofossil analyses show.

The hummocks which break up this pool layer are either long persistent features or, as at 540, 600 and 680 cm HS they are tussocks of *Eriophorum vaginatum* and associated *Calluna* which appear to have risen 'spontaneously' from the *Sphagnum* lawn just before pool formation – unless, of course, they were outgrowths of hummocks now cut away, hummocks which existed where the drainage ditch now is, and which grew sideways in the manner of the 550-500 cm HS hummock. In any event, at about the midway stage of infill of these upper pools there was a change of community and hence of peat type formed. A band of more humified (up to H6) peat appears, rich in *Calluna* and *Erica* (field determinations), and of a peculiar greenish-black colour. This seemed to be due to algal slime; it quickly disappeared from the peat kept dark in monolith tins. Although this peak is certainly more humified than, say, the ginger *Sphagnum* peat below it, it was difficult to assess the humification precisely. This was because the peat was a mixture of relatively fresh looking Sphagna and blackened *Calluna* and *Eriophorum*, with a fair amount of unidentifiable organic material; as is shown in the macrofossil analyses. It could be, of course, that when the peat was cut and exposed the humification process was still going on and that one cannot therefore precisely apply the von Post scale to such material. After this drier phase the bog again seems to have become wetter, for in a number of places (130-210 cm HS, 500-550 cm HS, etc.) fresh *Sphagnum* peat began accumulating over the *Calluna* rich bands until cut and drained in 1956.

Based simply on the field stratigraphy – as was most of Walker's work in 1961 – one is driven to the conclusion that bog growth at Bolton Fell Moss is in no way cyclic, *sensu* Osvald 1949, but rather that it is spasmodic and a response to shifts in the precipitation/evaporation ratio as well as a number of other factors. The persistence of hummocks and the flooding of large stretches of surface, both particularly well shown in the stratigraphy of section HI, cannot be accommodated within a cyclic theory. Further work on section HI – time-correlations across the section and changes in *Sphagnum* species, etc. – are reported in sections 5.3 and 5.4.

5.2.7 Sections HII-HVI

These five serial sections were cut at 20 and 40 cm intervals behind that part of section HI which shows an apparent cycle of hollows and hummocks – from 200 to 420 cm HS. The sections obtained are illustrated individually by standard diagrams and some by photographs (figures 21-26 and plates XVII and XVIII), and by specially drawn three-dimensional diagrams (plate XIX and XX) which give a simplified but, it is thought, useful visual impression of the stratigraphy. The decision to investigate part of the section in this way was prompted by interest in the fate of the two upper hummocks in the section, in the size and shape of the pool layers and in whether the spacing and occurrence of different types of community was regular or not. Monoliths were taken at various points from the faces, as indicated in the stratigraphic diagrams, but on reflection it was decided that the labour of full pollen and macrofossil analyses on these monoliths would not be repaid with much greater insight into the processes at work than had already been gained from the HI analyses. A few spot analyses of *Sphagnum* peat were done however, which confirmed the general succession seen in the HI analyses (section 5.4), and a pollen analysis on peat from the upper pool layer in section HIV proved it to be contemporaneous with the same pool from section HI. Besides recording

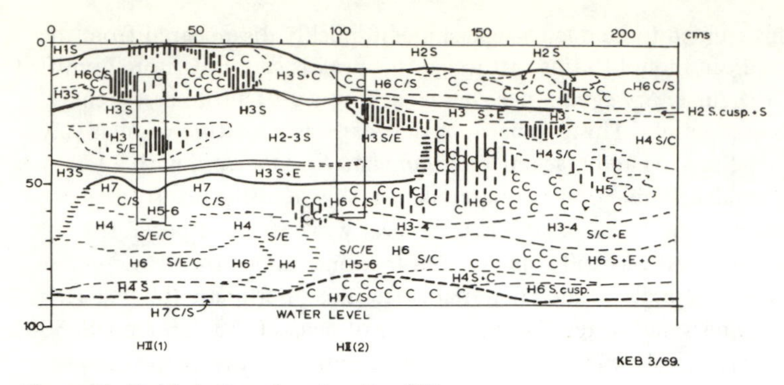

Figure 21. Field stratigraphy of section HII.

the stratigraphy of the peat faces, after an interval for weathering, the side walls of the pit were also recorded – and give a picture of development which is in some ways easier to comprehend than the serial sections (see figure 26).

Section HII, only 20 cm behind HI, shows, as expected, a basic similarity to the first section (figure 21). The middle pool mud (0-100 cm HS), again overlying a humified surface, is one of the most striking features (plate XVII). It is again separated from the humified surface by a few centimetres of fresh *Sphagnum* peat containing *S. cuspidatum.* The lower pools are not represented by pool muds – presumably they were shallowing out (see HIII stratigraphy) – but there is a band of *Sphagnum cuspidatum* peat and fresher H4 peat to the lower right of the diagram and two bands of H4 peat to the lower left, both of which might correlate with the HI pools. (Simple correlation by depth alone is not reliable due to shrinkage of the exposed peat, changing water levels in the ditch, etc.). The right hand hummock of section HI is still prominent, though a lens of H3-4 *Sphagnum-Calluna* peat seems to be encircling humified

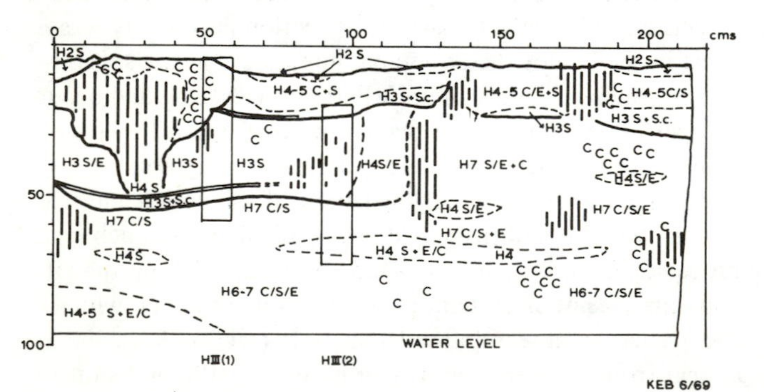

Figure 22. Field stratigraphy of section HIII.

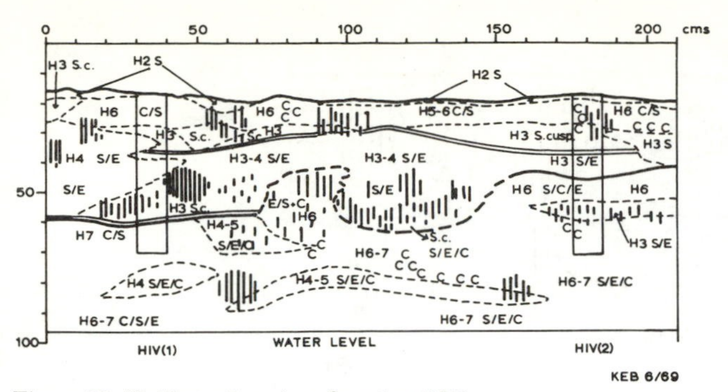

Figure 23. Field stratigraphy of section HIV.

peat of section HI at a depth of c.55 cm. The left hand hummock appears in HII merely as a slightly more humified lens of peat, richer in *Eriophorum* than the surrounding slightly fresher peat and completely 'undercut' by the pool mud (cf. diagram DI figure 17). The upper pool mud is now in place above the right hand hummock, separated from it by a few centimetres of fresh *Sphagnum* peat, and the uppermost 10-15 cm show a similar pattern to that already described from face HI.

Section HIII, again 20 cm behind HII, exhibits a greater amount of strongly humified peat than previously seen (figure 22) but still shares a number of basic features with HI and HII. Apart from a few lenses of H4 peat all the peat below about 50 cm VS is dark-brown H7 *Calluna-Sphagnum* peat and from 120-210 cm HS this humified mass extended almost up to the present surface. The middle pool is still a prominent feature and is succeeded by a block of H3 *Sphagnum* peat but now also has a prominent *Eriophorum vaginatum* hummock arising almost immediately above and possibly behind it. This hummock continued as an important structure up to the surface but elsewhere on section HIII the upper pool and the infill above it is present and similar to that in HI and HII.

The stratigraphy of section HIV (figure 23) also bears a likeness to the sections discussed above; it is again 20 cm behind the previous section. The upper and middle pools are still present, though one notices that the middle pool mud now lies directly on the humified peat, having sloped backwards in the section steadily from its position at HI – the diagram of the west side wall of the pit makes the relationships clear (figure 26). The generally humified lower part of the section is a direct continuation of the surface recorded in HI; it confirms the importance of the surface in the general stratigraphy of bog areas H and D. As can be seen from the photograph (plate XVII) there are no prominent hummock forms in section HIV, the broad swathe of generally humified peat below the middle pool being overgrown by fresh *Sphagnum* peat which then has a further pool layer superimposed on it at about 15 cm VS.

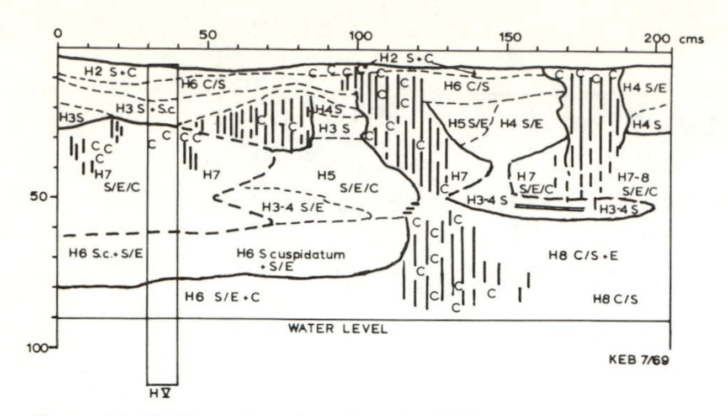

Figure 24. Field stratigraphy of section HV.

The familiar bands of H3 *Sphagnum* peat, H5-6 *Calluna-Sphagnum* peat and finally H2 *Sphagnum* peat then take the succession up to the surface.

In view of the similarities of HI, II, III and IV the next face was exposed 40 cm behind HIV rather than 20 cm. Peat accumulation in section HV (photograph, plate XVIII) began with a well humified peat, H6-8, with *Eriophorum* and *Calluna* common (figure 24). A concentration of *Calluna* and *Eriophorum* between 110 and 140 cm HS, 70 cm VS, probably represents the beginning of growth of a hummock, to the left of which the humified surface was flooded and overgrown by greasy *S. cuspidatum* peat, and to the right of which highly humified peat continued to accumulate. At 45 cm VS the hummock was almost overwhelmed by the growth of less humified peat associated with the formation of a small pool to the right but managed to survive as an entity and continue as a major hummock feature up to the surface. Above the small pool (150-175 cm HS) a drier *Sphagnum-Eriophorum-Calluna* community established itself, producing H7 peat, and acting as the base for a further *Eriophorum-Calluna* hummock, which also reached up to the surface. Between these two hummocks rather less humified peat accumulated, perhaps in unison with the peat accumulating to the right of the section (190-205 cm HS) and to the left, the H3 *Sphagnum* peat at 85-100 cm HS, 25 cm VS. Later during this period of rejuvenation the H7 peat which had accumulated over part of the lower pool and grown without much internal variation to form the large block shown in figures 24 and 26a was clearly flooded at 20 cm VS, 40 cm HS, to give a wet hollow with *Sphagnum cuspidatum.* This wet community apparently spread to the right, over the *Eriophorum* tussock which had sprung from the humified surface, ending against the persistent central hummock. Following this rejuvenation, which may perhaps be equated with the upper pool of sections HI-IV, comes the now familiar humified band of *Calluna-Sphagnum* peat, followed in turn by unhumified fresh Sphagna.

The final section in the series, HVI, was 40 cm behind HV; no further sections could be cut due to the stacking of peat immediately behind HVI. Not surprisingly the stratigraphy of HVI (figure 25) is rather different to HI, being much more humified on the whole (but especially in the upper 50 cm) – a full appreciation of the structure of this section would only be gained by excavating a longer length of ditch or excavating at right angles to the face; neither course of action was possible; so the following interpretation is to some extent imperfect.

The basal peat represents a generally dry plant community (humification 8), dominated by *Calluna* and *Eriophorum vaginatum*, with two areas of tussocky *Eriophorum* at 10-50 cm HS and 80-130 cm HS enclosing a small pool, 20 cm across and filled with *Sphagnum cuspidatum.* As the humified peat built up the pool migrated somewhat to the left before being infilled by moderately humified peat at 45 cm VS. Meanwhile, between 150 and 200 cm HS and at a depth of 63 cm an open water pool formed, laying down a remarkable thickness of pool muds – 9 cm in all. This may represent fairly deep water or some combination of chemical conditions favouring algal growths rather than *Sphagnum cuspidatum,* a possibility discussed by Boatman (1977). The mud is intercalated by a wedge of coarse *Sphagnum* and overgrown by a cymbifolian *Sphagnum-Eriophorum* which leads up to a drier community richer in *Calluna* and *Eriophorum,* and which seems to have formed a distinct surface at about 30 cm VS. It seems likely that this surface, representing perhaps a slow-down or cessation of peat growth for a few years, may have continued from the right hand side of the section at 20 cm VS, dipped down to 30 cm VS across to 100 cm HS then risen in some sort of hummock from between 100 and 60 cm HS, continued down across the H3-H5 junction between 60 and 30 cm HS to join the H8-H3 surface. It seems reasonable to postulate this in the light of the overgrowth of the right and left hand surfaces by a wet *Sphagnum* lawn, at the same depth and therefore probably at about the same time. In both hollows short pool muds formed at a later stage, followed by a fresh peat containing *Sphagnum cuspidatum.* The whole surface then became quite dry to judge

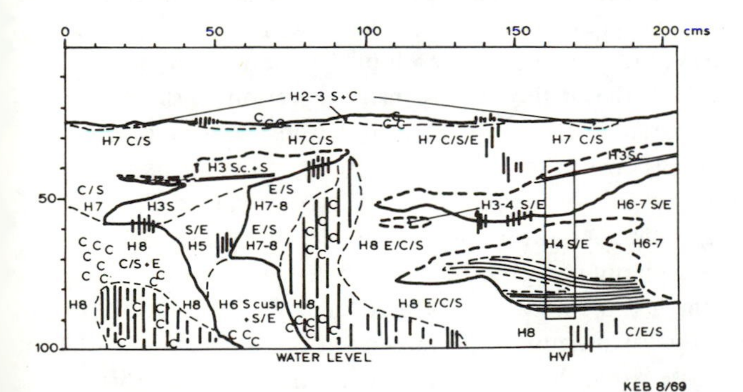

Figure 25. Field stratigraphy of section HVI.

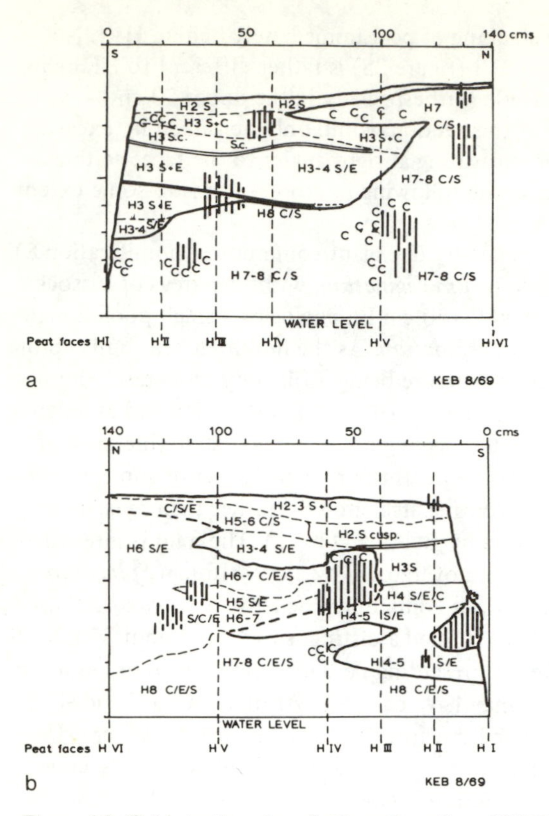

Figure 26. Field stratigraphy of side-wall sections HI-VI West and East.

from the strongly humified *Calluna-Sphagnum* peat, above which lie only a few centimetres of fresher *Sphagnum* peat.

Although the point has been made that this section exhibits a greater degree of humification than the rest of the area H sections, there are some similarities – one might, for example, equate the two pool levels with the middle and upper pools of HI-HIV, the difference in depth of the middle pool being accounted for by the fact that it slopes back into the section, as shown in the western side-wall stratigraphic diagram (figure 26a). Consideration of the stratigraphy of the side-walls does in fact clarify the status of the HVI stratigraphy. In figure 26a, of the west side of the pit,the humified peat can be clearly seen to rise up into a large hummock form; the stratigraphy of the east side of the pit (figure 26b), though more complex, tells essentially the same story. Such a large hummock complex, within which deep pools may exist (section HVI), puts one in mind of the 'island hummocks' described earlier (section 3.1.3), an island which in this case lasted the whole depth of section H (some 1500+ years). The west side-wall diagram and photograph (plate XVIII) show particularly well how the hummock formation maintained its integrity against the incursions of two floodings and the rapid accumulation of wet *Sphagnum* carpets to the south. The rate of accumulation of peat in hummocks must obviously play an important part in this maintenance and this factor will be discussed later.

Finally, with regard to the stratigraphy of HI-VI, it was thought that a simplification of the stratigraphic diagrams into block diagram form might be useful in giving an overall impression of the build-up of the section. Two methods were used, the first involving drawing up transparent serial sections, viewed from above and to the right of the sections. Plate XIX is the result and has some merit in bringing out firstly the simple number of pool muds present and, secondly, the persistence of a 'corridor' of pool/hollow through the central area of the block, flanked by hummocks. The second method involved the perspective drawing of the block, representing the curvature of hummocks by airbrush techniques, and making the block 'transparent' until, from the perspective of the viewer, a layer of humified (H5 and over) peat is encountered. This technique brings out the massive bulk of humified peat towards the rear of the block and helps one visualize the relative volumes of hollows and hummocks, and the position of the pools in space (plate XX).

5.3 POLLEN ANALYSES AND RADIOCARBON DATING

Correlation and dating of the various stratigraphic stages described in the previous section was performed according to the methods outlined in section 4. The results are presented here not in the order in which the analyses were performed but in what is thought to be a more logical sequence. The first analyses, on the monoliths from sections A and B, demonstrated the retardation at section A caused by drainage through the overlap of their pollen curves. The section C diagrams showed the artificial 'revertence' to forested conditions caused by the overturned layer of peat. Pollen diagrams from sections D and G were also in preparation at this time (1968-69) but all work on sections A-G was put aside once the overturned layer at section C had been discovered and during 1969-70 a master pollen diagram was constructed from section HI, monolith 9.

This monolith was chosen as it was known to be undisturbed (test counts) and because, passing through no less than five pool layers which formed in what looked like a very persistent stratigraphic hollow, there was less chance of a hiatus in peat accumulation or of widely different rates of accumulation. This master pollen diagram is therefore described first, along with the five radiocarbon dates on the same monolith. The pollen correlations of the other monoliths of section HI, numbers 1 to 8, with monolith 9, are then assessed in some detail to demonstrate the technique employed and the closeness of the correlations, followed by the pollen correlations of monoliths from other sections in alphabetical order and in a less detailed fashion.

5.3.1 Section HI – monolith 9

This pollen diagram (figure 27) was at first only counted down to the depth of the monoliths, that is from 0-105 cm. Because the counts from 95-105 cm clearly showed a phase of declining agriculture and increasing tree regeneration it was decided to explore the pollen analytical record further back in time, partly to put this agricultural phase in context and partly because it was thought possible that the lower stratigraphy of some sections (especially GI and GII) might be older than the 100 cm level in HI9. The record was therefore extended to 150 cm by making counts every 5 cm on a core extending from 100-150 cm. The diagram was then zoned for ease of discussion.

All of the HI9 diagram falls within Godwin zone 'VIII modern' or chronozone 'Flandrian III' of West (1970) so obviously no large-scale system such as these could be adopted. Consideration was given to using an extension of Walker's (1966) scheme for the Cumberland Lowland, but as use would only ha been made of his topmost four zones (C20-C23), and since even zone C23 does not extend up to the 1800 AD level with its characteristic pine increase, this zonation was rejected in favour of simply lettering the zones A to J, working upwards from the base. This sort of subdivision for localized or specialized usag has been employed by a number of workers, most recently by Atherden (1976) and Tinsley (1976).

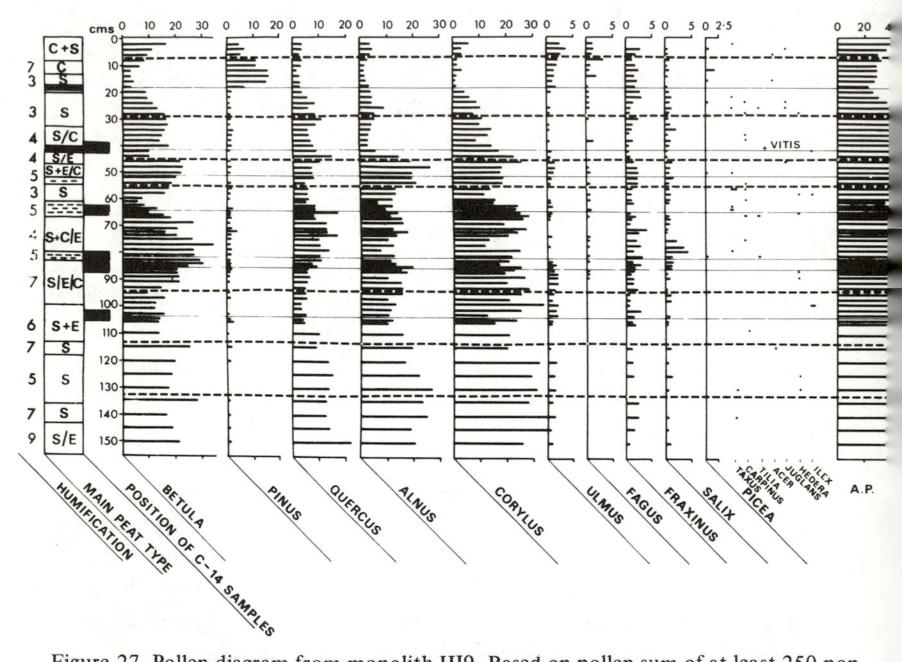

Figure 27. Pollen diagram from monolith HI9. Based on pollen sum of at least 250 non-mire pollen. Summary diagram showing main types only.

The features used in delimiting the zones are noted below; the radiocarbon chronology and land-use history correlations are discussed afterwards.

Zone A: 150-132.5 cm
This earliest zone is demarcated by the uniformly high AP (Arboreal Pollen) percentages (90 % of Non-Mire Pollen or more), very low grass pollen percentages (around 5 %) and the absence of any representation of cereals, including rye, or of *Rumex*. Percentages of other indicators of open ground or agriculture are also low or non-existent – for example, Compositae, *Artemisia*, Cruciferae and *Plantago* (around 2 %).

Zone B: 132.5-112.5 cm
Pollen of cereals and *Rumex* appear for the first time, along with an increase in *Plantago, Pteridium*, and Gramineae (up to 10 %). Tree pollen meanwhile declines throughout the zone from just over 90 to just under 80 %.

Zone C: 112.5-94 cm
Tree pollen declines below 60 % (and down to 37 % at 106.5 cm) and stays below this level throughout the zone with one exception (99 cm, AP = 64.2 %, largely due to *Corylus* at 33.6 %). Gramineae meanwhile climbs above 20 % for the first time and stays above that level throughout the zone, reaching a peak

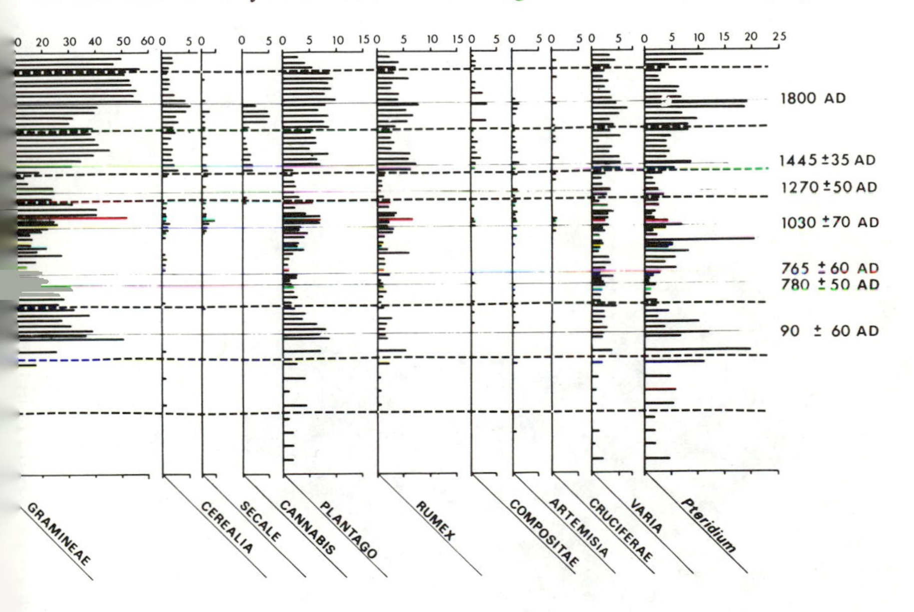

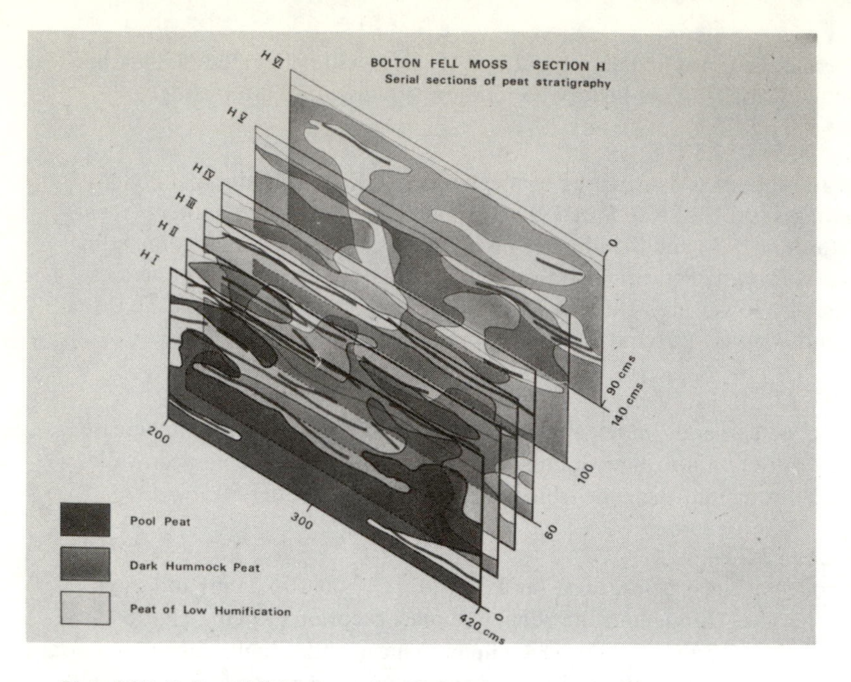

Plate XIX. Bolton Fell Moss: section H. Serial sections of peat stratigraphy.

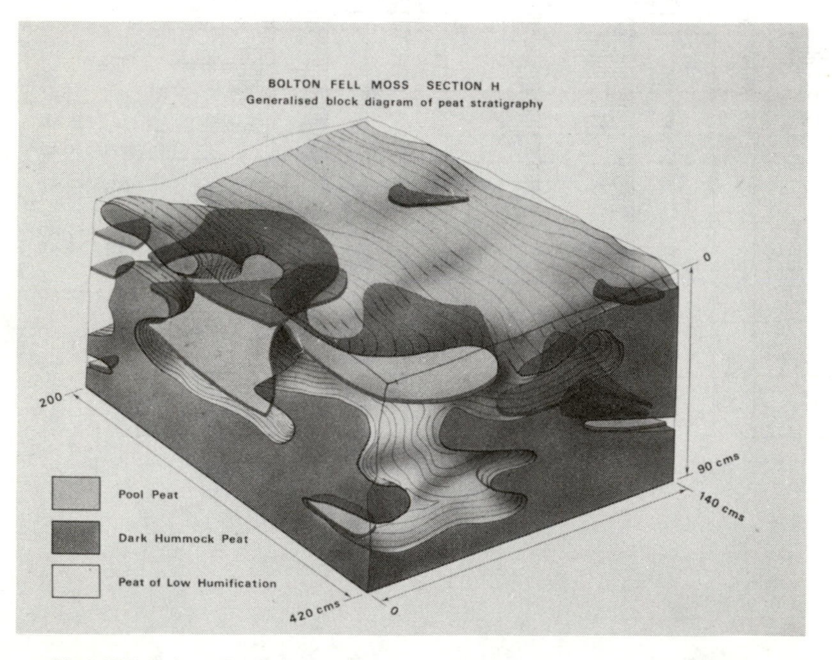

Plate XX. Bolton Fell Moss: section H. Generalised block diagram of peat stratigraphy.

of 50 %. *Plantago* and *Pteridium* respond in a like manner, the former reaching over 5 % for the first time, while *Rumex* and Varia (other herbs such as Umbelliferae, Ranunculaceae, *Urtica,* etc.) reach and sustain higher levels than in zones A and B. Amongst the trees the lower levels of *Quercus* and of *Betula* are noticeable.

Zone D: 94-64 cm

This zone is marked by a return of tree pollen values to levels in excess of 60 % of non-mire pollen, individual spectra reaching almost 80 %. This forest regeneration period starts with *Betula* rising from levels of 10-12 % up to a sustained 20 % with a peak of over 30 %. *Quercus* follows more slowly, as one would expect, but approximately doubles its average Zone C value – 5 % up to 10 %. The fluctuations of *Alnus* and *Corylus* are less important. Gramineae levels in Zone D are all below 30 %, many values lying below 20 %, and this is paralleled by the generally low level of *Plantago, Rumex* and other herbs. *Picea* and *Tilia* occur for the first time, both at 81 cm, a little after the beginning of a discontinuous curve of *Fagus.*

Zone E: 64-55 cm

This zone, though short, is quite distinct. It is marked by a rapid fall in tree pollen down to a low of 32 % and a concomitant rise of grass pollen to a peak of 53 %. The arboreal species mainly affected are *Betula* and *Corylus,* though all species show some decline. The six pollen spectra which make up the zone are also notable in showing a constant presence of three very 'discontinuous' species indicative of agriculture – Cerealia, *Secale* and *Artemisia,* as well as a peak of *Plantago* and *Rumex,* and the first appearance of *Cannabis.*

Zone F: 55-45 cm

Another short but distinct zone, though the reverse of Zone E, being one of marked forest regeneration with tree pollen reaching 80 % for two consecutive counts and grass pollen down to 10-15 %, the lowest level since Zone B. All arboreal species respond with higher levels in Zone F; Betula rises first from a low of 5 % in Zone E to 22 % and both *Alnus* and *Corylus* more or less double their representation. On the other hand cereals, other than a little rye, disappear until the very top of the zone, and *Plantago, Rumex* and other herbs fall to very low levels.

Zone G: 45-29 cm

This zone opens with a steep rise in the percentages of various agricultural indicators – cereals and *Rumex* respond first, along with an abrupt start of *Cannabis* representation as a continuous curve. Gramineae and *Plantago* respond next, grass pollen rising from 10 % to an average of over 30 % throughout the zone, and Compositae, *Artemisia,* Varia and *Pteridium* all show higher levels

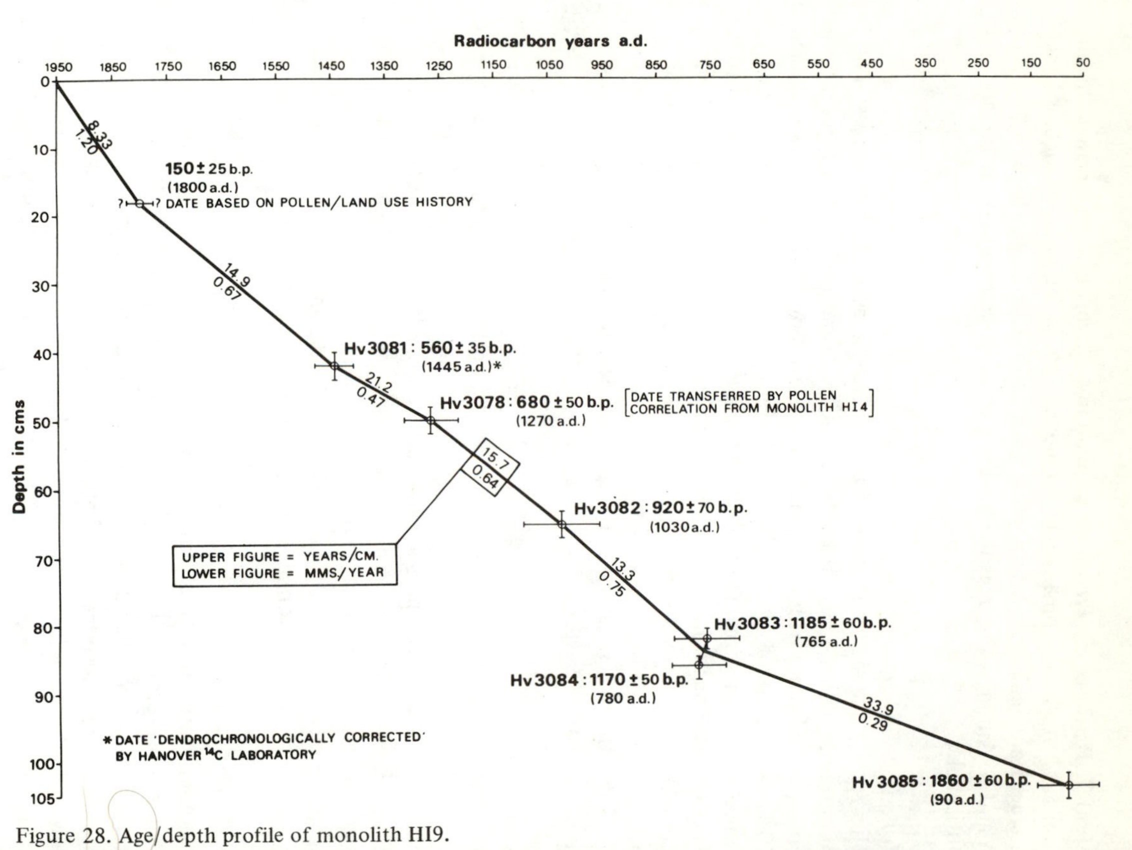

Figure 28. Age/depth profile of monolith HI9.

than previously. A single grain of *Vitis cf vinifera* was found at 41.5 cm. Arboreal pollen declines rapidly at the beginning of Zone G, down to an average of around 40 %. The main species affected was *Alnus,* which averaged around 20 % in Zone F but falls to an average of about 4 % in this zone. *Corylus* also declines generally, but *Quercus* and *Betula,* after an initial fall stabilize and even rise towards the end of the zone – at which time the agricultural indicators are also down from higher levels earlier in Zone G.

Zone H: 29-19 cm

This short zone is peculiar in exhibiting a fall in Gramineae pollen along with a general decline in arboreal pollen frequencies – a peculiarity accounted for by rising percentages of Cerealia and *Rumex,* high levels of *Plantago,* compositae and other herbs and especially by this being the 'Acme Zone' of *Cannabis* pollen. Grass pollen levels are, however, still high (30-40 %) relative to the diagram as a whole, and *Pteridium* spores exhibit a definite peak at the end of zone. Arboreal pollen falls from more than 40 % at the beginning of the zone down to about 25 % by the end, all species being affected except *Fraxinus,* which not unexpectedly rises.

Zone I: 19-7 cm

This zone is probably the clearest defined zone of all. Gramineae pollen leaps to more than 50 %, accompanied by a drastic fall in most tree pollen percentages (*Betula, Quercus, Alnus* and *Corylus* all down to around 5 % or less) but with a dramatic rise in pine pollen percentages (from less than 1 % up to 16 %). *Ulmus* and *Fagus* join *Pinus* with small though increased percentage representations. Meanwhile the percentages of cereals, *Rumex, Pteridium* and Varia, though down from their Zone H peaks, maintain a fair representation, and *Plantago* remains at a high level. Also important as 'negative' indicators are the absence of *Cannabis* and, slightly later, of *Artemisia.*

Zone J: 7-0 cm

This zone is demarcated from Zone I by the fall in the pollen of *Pinus* and *Plantago* and to a lesser extent Gramineae, and the rise of other arboreal percentages, particularly *Betula* but also *Alnus* and *Corylus.* The absence of *Secale, Cannabis* and *Artemisia* is also diagnostic of Zone J.

Correlation of these pollen assemblage zones with the land-use history of the area can now be considered.

The general picture given by the pollen evidence, of only two major impacts on the vegetation – Roman and Norse – before general clearance starting in the late Middle Ages, is in good agreement with the known history of this backward corner of England. The details of the land-use history/pollen correlations are given below. No detailed correlations with other pollen diagrams are considered, partially because of the local nature of much of the evidence, but one may note a similar pattern in the recent diagrams of Atherden (1976), Roberts *et al* (1973), Tinsley (1976) and Donaldson & Turner (1977), all from northern England.

Zone A. A largely forested landscape, without cereal cultivation on any scale but with clearings for pastoral activities, indicated by the sparse plantain and bracken pollen. This zone dates from 1440 BC to 840 BC by extrapolation of the peat accumulation rate between radiocarbon dates of 90 AD and 780/765 AD (the radiocarbon dates on monolith HI9, and the assumed straight-line rate of accumulation of peat between them, is shown on figure 28). Use of this slow rate of accumulation, 1 cm in 34 years, is justified by the high humification of peat below 100 cm. The zone is therefore Middle/Late Bronze Age and the low level of agricultural activity is in accord with the sparse archaeological evidence (Scott 1966, Collingwood 1933, Fox 1959) and with previous palynological work (Walker 1966, Pennington 1970).

Zone B. This period, from approximately 840 BC to 160 BC (again extrapolated from above using the 1 cm in 34 years accumulation rate) shows unequivocal evidence of settled agriculture in the presence of cereal pollen, which is accompanied by the beginning of the *Rumex* curve and a rising grass pollen curve paralleled by a falling tree pollen curve. Though most of the pollen zone falls in the general chronological limits for the Iron Age in Britain (circa 600-500 BC on) one must not think of northern Cumberland as being by any means in the mainstream of developments. The Lowland-Highland split, the cultural gradient from south-east to north-west of Fox (1932, 1959), is a very real factor in the history of the Cumberland Lowland. The sparse archaeological evidence from the area during these pre-Roman centuries and the lack of cultural innovation (Neolithic-Bronze Age cultural traditions and farming types remaining unaltered until Roman times) is noted by Cunliffe (1974) and by Feachem (1973), while Scott (1966), speaking of the area just over the present Scottish border says: 'It is not without reason that the later first millenium BC has been called the Dark Age of prehistory'. Pollen analytical evidence from the rest of north-west England supports this view – Smith (1959), Oldfield (1960) and Birks (1965) – and Walker (1966), working in the Cumberland lowland itself, concluded that the archaeological developments of the Bronze and Iron Ages (the latter . . . 'in the usual sense was hardly represented at all in Cumberland'), 'did not significantly influence the type of agriculture which had been practised in the lowland since middle Neolithic times'.*

If one then excludes improved agricultural techniques the explanation for the increasing farming impact of zone B must lie in increasing population or

* Davies & Turner (1979) come to a similar conclusion (p.789 and p.801).

in clearances being made nearer to Bolton Fell Moss or both. The soil map in the Geological Survey Memoir (Trotter & Hollingworth 1932) shows a division between 'Heavy Loam' and 'Clay' being dominant to the east of Bolton Fell with 'Light Loam' and various sands to the west and south. Slow penetration of small farming communities from the coastal areas therefore seems likely.

Zone C. This zone encompasses the Roman period, extending from 160 BC to 465 AD, with a radiocarbon date of 90 ± 60 AD (Hv. 3085). Palynological evidence of Roman agriculture is unfortunately rare, only one example, from her own work at Flanders Moss in the Forth Valley, being cited by Turner (1975). The evidence for the zone C clearance being Roman is, however, very good. For one thing the radiocarbon date was on a large, carefully sampled piece of humified acid peat, with no evidence of rootlet penetration or any other source of contamination, and the date obtained agreed, within the one standard deviation of 60 years, with the author's pre-assay estimation based on averaged accumulation rates. There is therefore little reason to doubt it – if the rate of peat accumulation between this date and the next two, 765 and 780 AD, seems rather slow at 1 cm every 34 years it must be noted that the peat concerned is of humification 6-7, compared with an average of humification 4 above the 770 AD level.

Secondly the land-use history evidence in favour of zone C being Roman rather than Bronze or Iron Age, or post-Roman, is overwhelming. Bolton Fell Moss is no less than 5 km north of Hadrian's Wall (built 122-130 AD) and is centrally placed in a rough rectangle formed by the 70 acre walled town of *Luguvallium* (Carlisle) and the fort at *Petriana* (Stanwix), 14 km to the south-west; the fort of *Uxellodunum* (Castlesteads) and its associated civil settlement only 6 km to the south; the fort of *Camboglanna* (Birdoswald) and its associated civil settlement and watermill, 12 km to the south-east; and the two 'out post' forts of *Castra Exploratorum* (Netherby), 9 km to the west, and *Banna* (Bewcastle), 9 km to the east (Ordnance Survey – Map of Hadrian's Wall, 1964, and Map of Roman Britain, 3rd Edition 1956). These forts and the Roman walled town of *Luguvallium*, were the 'pivot' of the whole system, for it was mainly from south-west Scotland that trouble was expected (Wilson 1967). Even if one takes quite low figures for the settled native population, the garrison of around 15,000 men (Richmond 1958, Birley 1963) for the wall area must have made a great impact simply in terms of wood for fuel and building material, to say nothing of scrub and woodland clearance for military purposes – that is, to clear a 'no-go' zone both to the north and to the south of the wall, the latter clearly connected with the vallum (Wilson 1967). This activity began with the major campaign of Agricola in 78-84 AD and the pre-Wall forts such as *Vindolanda* (circa 80-125 AD, Birley 1973), followed by the vast job of wall and fort construction (122-130 AD). Thus the deforestation and great increase in grasses and weeds dated to 90 ± 60 AD (a date not greatly affected by re-calibrations of the C^{14} time scale – see McKerrell (1975)), is in excellent agreement with this human impact on the environment.

After the peak of clearance at 106.5 cm there is some revertence to more forested conditions through the rest of the zone and little evidence of arable agriculture, save for the presence of *Artemisia* and Compositae, but the general character of the landscape would still be a fairly open one to judge from, for example, the level of Gramineae pollen. This period, extending to about 465 AD at the end of Zone C, was one of fluctuating frontiers and periodic rebuildings of the Wall following native revolts in 196-7 AD, 296 AD and 367 AD (Hunter-Blair 1963, Birley 1963). In between times life must have been fairly settled to judge from the extensive civilian settlements outside many forts and the extension of farming necessary to feed civilians and soldiers is reflected in the pollen curves – previous erroneous ideas of supplying the garrisons from the south have recently been decisively rejected by Manning (1975) who draws our attention to the cultivation terraces south of Housesteads fort, one of the bleakest sites on the Wall, and concludes that: 'We should accept that a considerable part of the requirements of the military . . . was locally produced with all that involves for the expansion of agriculture and its effect on the landscape', a view hinted at by Walker (1966, p.201). The reconstruction drawings by Alan Sorrell in Garlick (1972) and Birley (1963) are therefore probably fairly accurate in their depiction of a semi-open landscape; further evidence is reviewed by Seaward (1976). Finally one must note that this view is contrary to some previous archaeological opinion, particularly that of Collingwood (1933) – '. . . disturbance was very small . . . almost at the vanishing point in the scale of Romanisation' (p.192), a view reiterated by Hogg (1972), but that it is in tune with other work mentioned above and with the evidence of the recent aerial survey results of Higham & Jones (1976).

The end of Zone C, drawn where Arboreal Pollen again consistently exceeds 60 % of non-mire pollen, is calculated to fall at around 465 AD, that is after the final abandonment of the wall around 400 AD and at a time when the romanized cities were falling to the Saxon invaders.

Zone D. This zone is calculated to lie between 465 AD and 1050 AD, and is a long period of subdued agricultural activity, arboreal pollen counting for some 70 % of the non-mire pollen and grass pollen often below 20 %. The early part of the zone, between 465 AD and c.770 AD exhibited higher grass pollen frequencies and this more open landscape tallies with the evidence for a peaceful withdrawals of the Romans early in the 5th century and the continuation of town life in, for example, Carlisle (Blair 1956, 1963). A small British kingdom Reget (Rheged), allied to Strathclyde and to the North Welsh, is thought to have existed covering southern Galloway and the Carlisle plain, the three forming an area of British (Celtic) dominance down the west coast (Ordnance Survey 1966). From 383 AD until c.600 AD, however, the history of this area is 'almost unknown', though the kingdom of Strathclyde is thought to have existed as early as 450 AD (Blair 1956). The Anglian kingdoms of Bernicia and Deira united to form Northumbria circa 603 AD, pushed west through the Tyne Gap probably settling at Brampton and Irthington amongst other places, and reached Carlisle sometime before Cuthbert's visit in 685 AD (Blair 1956). Settlement seems to have been on a small scale in what was again a border area, this time between Strathclyde and Northumbria, until the settlement of the Norsemen during the early 10th century. There is no evidence, pollen-analytical or archaeological, for the major clearances claimed to date from the 5th century by other authors – see Turner (1970).

The radiocarbon dates of 780 ± 50 AD and 765 ± 60 AD mark the major changes in humification in the peat of monolith 9 and this change to a wetter and/or colder climate can hardly have encouraged settlement, although some cereal pollen is present immediately afterwards, as is a single grain of spruce, *Picea abies.* Both possibly are due to the earliest Norse settlement of the area and the introduction of spruce, the commonest tree of the Norse homeland, some centuries before the earliest record (circa 1500 AD, Mitchell 1974), is a noteworthy event. The generally low level of agriculture remains until the very end of the zone, when rye and other cereals start to show continous curves, circa 1000 AD. By this date the lowland to the south-west of Bolton Fell Moss was quite well settled, to judge from place-name evidence (Ordnance Survey 1973) although some authorities believe Carlisle abandoned 800-1000 AD (Hogg 1972), and the zone ends circa 1050 AD, a date supported by the radiocarbon result of 1030 ± 70 AD, which marks a phase of wetter peat accumulation.

Zone E. This short zone of only some 140 years (1050-1190 AD) is marked by a rapid deforestation, with grass pollen values reaching 50 % at their peak for the first time in 1000 years; arboreal pollen is below 60 % during the zone and is the main demarcating factor. The main tree pollens affected, *Betula* and *Corylus,* and to a lesser extent *Fraxinus* and *Quercus,* seem to indicate that secondary, scrubby woodland was being cleared. This is followed by an almost equally rapid decline in agricultural activity and forest regeneration. The history of this period is somewhat confused (Ferguson 1890, Millward & Robinson 1972), the area being at times under Norman rule and for the rest a part of the semi-independent Scottish kingdom of Strathclyde or Cumbria, which stretched to the Rere Cross on Stainmore (Barrow 1956). It was, to quote this author (p.127) 'a thoroughgoing mixture of Briton, Northumbrian, Scandinavian and Gael', and Scottish kings ruled over Cumberland until 1092 AD when William II took Carlisle, built a castle and planted at least seven villages (Millward & Robinson 1972). The area was lost to the Scots again between 1135 and 1157 AD, due to civil war, Henry II retrieving the area for England at the latter date. The present border dates from more or less this time. Zone E then marks a period of some settled agriculture under rulers independent of the Norman conquerors, which, of course, explains the absence of Cumberland from the Domesday Book (1086 AD) and the survival of agriculture, including monastic influences such as Holm Cultram (Cistercian, 1150 AD) and Lanercost (Augustinian 1160 AD), in a period when other pollen diagrams from the southern Lake District record a brief forest regeneration due to the punitive expedition of William I in 1068-9 (Oldfield 1963, 1969).

Zone F. This zone is the last extensive regeneration of woodland (60-80 % non-mire pollen) and the last break in the cereal curve before the present day. Settled agriculture seems to have stopped almost completely during the zone, save for some pastoral activity. There is virtually no cereal pollen, and hemp cultivation, which had started to show at the zone E/F border, ceases, and all herb pollens sink to very low levels. All arboreal species benefit and the grass curve sinks to just over 10 % – pre-Roman levels. The period may be dated from the depth-age curve and three closely associated radiocarbon dates (1030, 1270 and 1445 AD, the central one transferred from monolith 4 by virtue of a near-perfect pollen correlation) to between 1190 and 1380 AD.

Although there were increases in population and expansions of settlements during this period, especially during the first part and especially due to the monastic foundations (e.g. the charter for the weekly market and four sheep and cattle markets at the already established town of Brampton dates from 1252 AD – Millward & Robinson 1972), the general character of the period was one of war, pestilence and famine, particularly during the 1300's. The 1200's, though unsettled, were not a time of major Scots incursions and the harvests in general were still good due to the medieval climatic optimum, but the 14th century 'was the most miserable that . . . the men of Cumberland ever had to endure' (Ferguson 1890). Largely because of the Anglo-Scottish Wars, beginning in 1296, we have the contrast of the 'citizens of Carlisle waxing fat on the wages of the soldiers and the money of the courtiers, while the wretched peasants around them starved' (Ferguson 1890), – or else joined the armies of one side or another and left the land uncultivated.

The diocese of Carlisle in 1291 was valued at £3171; in 1318 it was only £480. Over the same period, Lanercost's valuation dropped from £73 to nil (Millward & Robinson 1972, Ferguson 1890). Besides small raids too numerous to be recorded, major Scots incursions and serious devastation – often in response to treatment of the same kind from the English – are recorded in 1296, 1297, 1311, 1314, 1315 (Siege of Carlisle, surrounding country devastated by Scots for supplies, etc.), 1322, 1333, 1337, 1346, 1380, 1385 and 1387 (Ferguson 1890). The rich lands and treasures of Lanercost Priory were often the target for the Scots if Carlisle proved obdurate and the Priory was burnt in 1296-7 and in 1345 – from whence it never recovered. Such attacks on a potential centre of agricultural enterprise – and on Holm Cultram – were especially harmful to the rural economy but they were not the only factors at work. The well-known harvest disasters of 1315, 1316 and 1317 due to the weather (the early 14th century ushered in the Little Ice Age – Lamb 1966), and the Black Death of 1348-50, combined with the wars to give the dislocated agriculture apparent from the Zone F pollen statistics. The course and the effects of the plague have been fully described by Ziegler (1969); he points out that the main pandemic of 1348-50 was followed by outbreaks in 1361, 1368-9, 1371, 1375, 1390 and 1405 and that the Black Death alone was responsible for the death of 25-30 % of the population. On Cumberland he gives a graphic account of the agricultural disruption which is worthy of reproduction:

Of Cumberland still less is known. In this county, the diocesan registers are lacking; a tribute not so much to the Black Death as to the havoc wrought by the invading Scots. But though the Scots prepared the ground it was the plague which finally dislocated the agricultural economy. The accounts of Richard de Denton, former Vice-Sheriff, presented for audit in 1354, show vividly what damage had been done. Because of the 'mortal pestilence lately raging in those parts', he reported, 'the greater parts of the manor lands attached to the King's Castle at Carlisle' were still lying uncultivated. For eighteen months after the end of the plague, indeed, the entire estate had been let go to waste 'for lack of labourers and divers tenants. Mills, fishing, pastures and meadow lands could not be let during that time for want of tenants willing to take the farms of those who died in the said plague'. The jury found that Richard de Denton had proved his facts and accepted the greatly reduced value of the estate (Gasquet pp.183-184). The city of Carlisle was relieved of many of its taxes in 1352 because 'it is rendered void and, more than usual, is depressed by the mortal pestilence'.'
(From Ziegler 1969, p.185)

The Peasants Revolt of 1381 and the widespread agricultural depression all over Europe (Myers 1963, Slicher van Bath 1963) complete the picture of decline from the agricultural zenith of the early Middle Ages, a decline which Slicher van Bath describes as 'The Great Death', lasting from 1300 until revival began in about 1450.

The land use history of the period calculated to be covered by Zone F, 1200-1400 AD, therefore fits extremely well with the pollen record giving one confidence in the dating of this important correlative zone.

Zone G. A remarkable revival in agricultural activity takes place in this zone, despite a worsening climate (section 5.5) and a still far from settled border situation. Alder and birch were particularly affected, probably indicating clearances on wetter land around the bog, despite the higher water tables apparent from the widespread pool layer dated at 1445 ± 35 AD. After the sharp woodland decline at the beginning of the zone the arboreal pollen total stays more or less constant at 40 %, and in the irregular fluctuations of the hazel curve there is some suggestion of the practice of hazel coppicing, possibly with standard oak preservation. Ferguson (1890) notes that 'there is little of distinction to record of Cumberland in the fifteenth century, or the early sixteenth century . . . the episcopal registers are missing' (p.237-8). Carlisle was beseiged in 1461 during the Wars of the Roses but in general the Border troubles were either reduced by the diplomacy of Henry VII (1485-1509), or moved to the east with, for example, the Battle of Flodden, 1513. The division of the Debateable Lands, between the rivers Esk and Sark, by the 'Scotch Dyke' in 1552, following the Scottish defeat of 1542 at Solway Moss, further defused the situation.

However the century of 1503 to 1603 is that of the heyday of the Border reivers or moss-troopers, popularly thought to have been constantly raiding and feuding. While some of the decline in cereal and herb percentages later in zone G may be put down to this, the recent detailed study of the Border reivers by Fraser (1971) goes some way to reconciling the paradox between an unsettled warring century and fairly high levels of cereal production and low levels of woodland cover. Two main points stand out from Fraser's researches; firstly that there was a definite raiding 'season' of autumn to spring when the cattle and their owners were back from the sheilings or summer grazing grounds high in the Border hills. The most favoured time was *after* the harvest of (mainly) barley and oats, from Michaelmas (29 September) to Martinmas (11 November) for 'then are the fells good and drie and cattle strong to dryve' (Fraser 1971, p.71). Naturally there were raids at other times but the needs of producing a crop for food and for brewing affected both sides of the Border and the existence of an autumn-winter season is further borne out by the fact that a Border guard system drawn up in the 1550's operated day and night but only from October to mid-March (Fraser 1971, p.72, also p.35 and pages 16-18 for confirmation of the devastation of the 1300's).

Secondly it seems likely that, despite the raiding, the emergence of strong 'clans' – the 'riding surnames' – such as the Armstrongs, Bells, Grahams and Hetheringtons (the last being centred on Hethersgill, immediately south-west of Bolton Fell Moss) gave some degree of permanence to the agricultural settlements of the area, thus stimulating agriculture – but, it must be emphasised, only in a relative sense. The Union of the Crowns in 1601 and local events such as the Barony of Gilsland Survey, 1604, mark the end of some 15 centuries of warfare and unsettlement.

Zone H. This zone is calculated to lie between about 1640 AD and 1785 AD and exhibits an overall decline in woodland cover – except for ash trees, a common hedgerow species today – and a general increase in all agricultural indicators, arable and pastoral. There is an initial fall in grass pollen values, mainly due to the relative increases in *Cannabis, Rumex* and Compositae – the residual grass values allied with the very slight increase in arboreal pollen, perhaps being due to the Civil War. Carlisle was under seige by Roundhead forces from October 1644 until June 1645 and the surrounding countryside scoured for provisions for both forces, and Jones (1956) comments upon the relative poverty of Cumberland and Westmorland in the late 17th century. The Jacobite risings of 1715 and 1745, though involving Cumberland quite heavily, were over too quickly to leave any great impact on the countryside (Ferguson 1890), and the rising tide of agriculture during the 18th century is well shown in the later part of Zone H.

One must not assume that the 'New Husbandry' of Arthur Young and others had an immediate impact in the North; as with most innovatory practises progress was slow (Slicher van Bath 1963). The upsurge in hemp growing, for rope-making, was probably a direct response to the Industrial Revolution, touching Carlisle as elsewhere with cloth making starting up in the 1740's (Ferguson 1896). Agricultural improvements were being made in north Cumberland during the 1700's – a parcel of land (100 acres) right on the south-east side of Bolton Fell Moss was enclosed by private agreement in 1735 (Carlisle Record Office reference no. D/Ha/2/16 and see figure 11) and to the west of the moss some 4,200 acres were enclosed in the 1730's. The pace of such improvement and enclosure quickened as the century wore on (Elliot 1973, George 1953), and the general trends of such improvements in grassland cultivation, without specific reference to Cumberland, have been comprehensively covered by Fussell (1964) and by Franklin (1953).

Zone I. This zone sees the beginning of the creation of the modern landscape, stimulated by the well-known agricultural 'boom' of the Napoleonic Wars (Thomson 1950) and the country estate 'movement', led in the north Cumbrian area by innovators such as Sir James Graham of Netherby Hall (Spring 1955). The clearing of scrub is a marked feature of the zone – birch, alder and hazel are reduced to almost zero pollen representation. Oak is also reduced and in its place we see plantations and roadside strips and windbreaks of pine, elm, beech and ash and approximately 2 % of the spruce pollen at one point. These together account for a slow rise in the arboreal pollen total.

There is a dramatic rise of grass pollen to a sustained representation of over 50 % for the whole zone and non-rye cereals reach a peak at the lower zonal boundary, as do *Rumex,* Compositae and *Pteridium.* These changes all relate to fairly massive changes in agricultural intensity and in technique – the cultivation of hemp ceases dramatically while that of rye ceases also and *Artemisia* disappears as a component of the weed flora.

The correlation of the above changes may be considered even more certain than those of previous zones. The late 18th century and early 19th century saw the start of what may conveniently be called the 'county' publications – volumes on the topography, antiquities, family histories and agriculture of the various counties of England, a movement related to that of the establishment of country gentlemen's estates. In Cumberland these publications begin with Nicolson & Burns 'History and Antiquities of Cumberland and Westmorland' in 1777, followed by Hutchinson's massive work (1794); Housman's 'topographical description' of 1800 (much of 'Housmans Notes' appear in Hutchinson's work); Jollie's guide and directory (1811), Pinnock's county history (1822), Jeffersons History (1838) and a number of other minor guides and directories. These are often tedious to wade through in search of information on agriculture; the author has found that only Housman's notes in Hutchinson (1794) and to a lesser extent, Nicolson & Burn (1777), are of any value. In the former publication one gains the distinct impression of a landscape undergoing great changes, with notes added in the press to the effect that 'much improvement has been made of late', and also of a landscape almost totally cleared – the amount of woodland in every parish was assessed by Housman and the amounts in the two parishes around Bolton Fell Moss – Stapleton to the north and east, Kirklinton to the south and west, the boundary running across the moss from west to east – are noted as follows:

Stapleton: Upon the hedges are a few trees, but scarce any regular woods (p.561, Vol. II). *Kirklinton:* Upon the banks grows a quantity of oak-wood in some parts; there are also upon the ancient hedges some few trees of ash, oak, birch etc, with two or three small plantations of fir upon the cultivated moor, but do not seem to thrive well. Hedges are good or bad according to the quality of the ground (p.570, Vol. II).

Similarly Scaleby parish, to the south-west, has 'no regular woods', nor have the parishes of Crosby, Stanwix or many others in the district. However, in those areas adjacent to the Netherby estate of the Grahams we find that 'Dr Graham planted a considerable quantity of fir (i.e. Scots pine), birch etc', and again: 'Sir James Graham planted, lately, about 30 acres with fir' (Hutchinson 1794). By 1845 the total plantations on the Netherby estate stood at 2300 acres (Spring 1955).

The Grahams owned a huge tract of land comprising almost the whole of the parishes of Kirkandrews, Nicholforest and Arthuret, an irregular area of about 11 km by 19 km only a few kilometres to the west of Bolton Fell Moss. Their great exertions agriculturally, particularly between about 1780 and 1850, fully documented by Spring (1955) transformed the local landscape and they had their imitators over a wide area of northern Cumberland. It is therefore to this period of the 19th century that the changes apparent in zone I must be referred. The date of 1800 AD for the depth of 18 cm is based on these considerations, particularly the rise in pine pollen; the farming boom of the Napoleonic Wars; and the demise of *Cannabis* due to importation from British India – 'Flax and hemp are now rarely seen growing here, though 50 years ago, a little hemp was sown by almost every cottager and statesman' – Bailey & Culley (1805, p.313), referring to Westmorland but there can be little doubt that the same applied to Cumberland. *Artemisia* is thought to have died out due to the introduction of deep ploughing methods early in the 19th century (Birks

1965) and all these indicators taken together must mean that the 18 cm level is dated to 1800 AD ± 25 years.

Further evidence, both of dating and of the extent and character of agriculture, come from the evidence of the enclosure of common and open fields and from agricultural reports. These reports include those commissioned by the Board of Agriculture, and various papers in the Journal of the Royal Agricultural Society, of which Sir James Graham was a founder member in 1840. Elliot (1956, 1959 and 1973) goes into great detail of both enclosure and improvements, detail which is not relevant here except to note that whereas the enclosure of 'common waste' by Act of Parliament averaged about 10-15,000 acres per decade from 1760 to 1800, in the decade beginning 1810 slightly over 80,000 acres were enclosed, followed by 60,000 acres in 1820-30 and then back to about 10,000 acres per decade from 1830-1880. Clearly, this is reflected in the grass pollen percentages of zone I, as well as in the extremely low levels of birch, alder and hazel pollen.

The two best known agricultural reports on Cumberland also date from the period covered by zone I (1785-1895 AD). The first of these by Bailey & Culley (1805) has a lot of detail on agricultural practices, which are generally considered rather backward, as are the houses of the inhabitants of the Bolton Fell area (p.207, '. . . of mud or clay . . . a miserable contrast to the . . . rest of the county') – all this was, of course, improved by the changing economic situation and the activities of the Netherby Estate, as already mentioned. The chief crops were then barley and oats, wheat being a fairly modern introduction and, on the sandy areas of the Brampton kame belt particularly, turnips and potatoes were grown with success – the former having been introduced only in 1755.

Later, in 1852, Dickinson gives a detailed account of the agriculture in a Royal Agricultural Society Prize Essay. The same crops as mentioned above predominate (in what would equate with mid-Zone I) but oats has lost ground and wheat seems to have gained in acreage – which would account for some of the fall in the cereal curve in zone I, modern wheat releasing very little pollen compared with oats and rye. The latter grain is mentioned as being 'very little grown' and then only on very sandy or peaty soils. *Plantago lanceolata* ('rib-grass') was now sown as part of a 'permanent pasture' seed mixture, along with the newly appreciated clovers, and it is interesting that *Trifolium,* though entomophilous, was noted amongst the 'Varia' pollen in Zone I. Of woodland Dickinson notes (p.281) that 'Every landed gentlemen's seat has its scores of acres . . . some estates have hundreds of acres . . .' mainly of 'Scotch fir' but also larch (no pollen discovered) and spruce, and he notes how much more woodland there is since the start of the great enclosure movement (1790). Again these descriptions fit very well with the pollen spectra of Zone I.

Finally, as the most accurate accounts of mid-19th century agriculture, mention must be made of the Tithe Returns and of the first Ordnance Survey maps and the survey books. The accuracy and usefulness of these sources has been reviewed by, amongst others, Grigg (1967) and Harley (1963, 1972). The Tithe Returns, that is the survey that was necessary to carry out the Tithe Commutation Act of 1836, have been studied for Cumbria by Bainbridge (1943) and the most striking feature is the relatively large amount of arable land recorded for the 1839-45 period. For example Hayton Parish, to the south of Bolton Fell, was 83 % arable for 1841. This is an exceptionally high figure partially explained by Hayton's position on favourable light soils closer to the market of Carlisle, but even Bewcastle, a northern hilly parish with 30,000 acres of titheable land in 1839, had 4,040 acres of arable, against 17,000 acres pasture and 268 acres of woodland – the remainder being presumably rough grazing.

The Ordnance Survey 'Parish Area Books' were produced between 1855 and 1886 and give to three decimal places the acreage of every enclosed parcel numbered on the 25 ″ plans (Harley 1963). More importantly they also give the use to which the land was being put at the time under broad categories such as 'Arable', 'Pasture', 'Wood', 'Houses', etc. For the parish of Kirklinton, immediately west of Bolton Fell Moss, these statistics occupy 25 double-columned pages and the Township of Hethersgill – that third of the parish immediately abutting onto the moss – has no less than 979 entries. Totalling up all the 'arable' entries gave a figure of 2,585 acres out of a parish acreage of 5,360 acres – 48 % arable cultivation. Woodlands accounted for only 4.5 % (not near any great estate).

This intensity of cultivation appears to be maintained throughout zone I, and although the arable crops do not of course show up as such on the pollen diagram, except in a great number of taxa being recorded, the 'golden age' of British agriculture is amply demonstrated by the extremely low percentages of non-plantation trees and the high levels of agricultural and open ground indicator species.

Zone J. This last zone, calculated as beginning in 1895 AD and continuing up to about 1940 contains only three pollen spectra. The surface sample was not included as it almost certainly contained old peat from that stacked up to dry. The surface in any case was drained in 1956 and the top of the profile was arbitrarily taken as 1950. Modern pollen-rain counts do not show a great divergence from the topmost spectrum of this diagram, at 2 cm.

The resurgence of birch and bracken, and to a lesser extent hazel and alder, and the fall of pine and of plantains can be related in a general way to recent agricultural history. The two World Wars had a dramatic effect on the amount of tillage (both for arable and temporary grass) and, of course, on woodland and waste land. Best & Coppock (1962) and Coppock (1971) give national surveys of these trends and show that in Cumbria tillage decreased by 50 % overall between 1875 and 1938, and in northern Cumberland the figure was almost certainly higher; local farmers remember the 1930's as a period of much hardship, with a lot of land around Bolton Fell Moss lying abandoned. The First World War did not have as great an impact as the Second in terms of 'ploughing-up', due to different government policies and far less threat to imports by submarine warfare, though the effect on the small percentage of good woodland was very great (Ryle 1969). The fall of pine percentages from 12 to 4 % between 8 and 6 cm on the pollen diagram can almost certainly be put down to this cause. At the same time (6 cm) there is a single peak of grass pollen – probably an increase in temporary grassland to meet war needs, but at the same time plantain values fall (not perhaps sown as part of the seed mixture, and/or, as noted by Franklin (1953), because cows strongly prefer plantain, even to grass or clover, and will 'eat it off close to the ground . . . before moving elsewhere'), and continue to do so into the agricultural depression of the 1930's. The increases of birch and bracken are in accord with this.

This master pollen diagram, HI9, therefore clearly shows a number of well-marked zones, often with abrupt changes of percentage value and/or of taxa; the dating of these episodes by means of a series of five internally consistent radiocarbon dates and two other dates fixed by pollen correlation and a major land-use event, has been shown to be in close agreement with the changes in the vegetation of the area, deduced from archaeological and documentary records, since Bronze Age times. The diagram can now be used to correlate the shorter diagrams from other monoliths and sections to provide a chronology of the peat stratigraphy.

5.3.2 Section HI – monoliths 1-8

Nine supplementary pollen diagrams were done from monoliths 1, 2, 3, 3a, 4, 5a, 6, 7 and 8 (shown on the stratigraphic diagram, figure 20) and correlated with the master diagram from monolith 9. Radiocarbon dates were obtained on monolith 4 (3 dates) and on monolith 6 (1 date), and for this reason rather more pollen analyses were performed on those monoliths. It was found, with experience, that the very close counting done across boundaries on monoliths 3, 4 and 6 was superfluous and that because of the marked changes in the diagram from monolith 9, and because the correlation was based on 22 curves at any one level, then individual 'spot' counts, especially on 'datum lines' such as pools, would suffice. With many correlations the best match was found by placing the 'unknowns' (that is, counts from monoliths other than HI9) between two adjacent spectra of diagram HI9. This means that in the correlations which follow an 'unknown' count is correlated at a depth from which there may be no actual count in diagram HI9, but at a level where it is equal to a percentage mid-way between two HI9 counts. Naturally the greater the number of pollen spectra the finer the correlation becomes, but it is hoped that the line has been drawn at about the right level and sufficient counts done

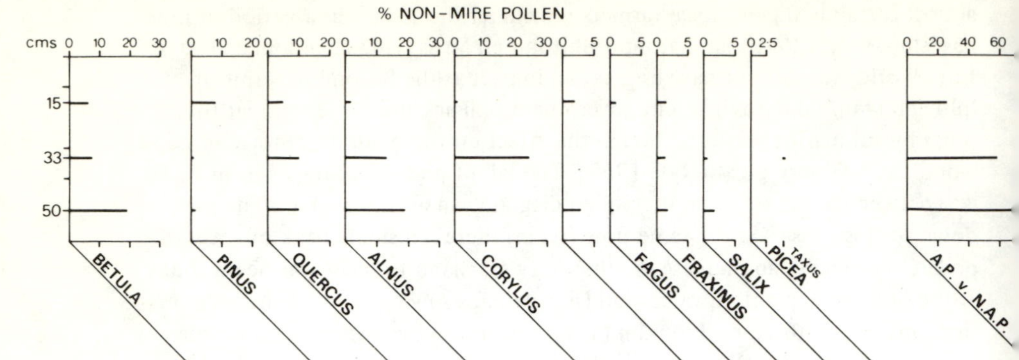

Figure 29. Pollen diagram from monolith HI1. Based on pollen sum of at least 250 non-mire pollen. Summary diagram showing main types only.

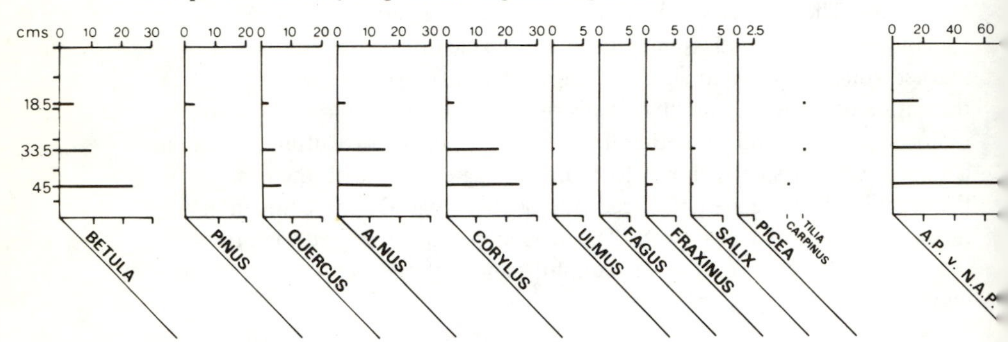

Figure 30. Pollen diagram from monolith HI2. Based on pollen sum of at least 250 non-mire pollen. Summary diagram showing main types only.

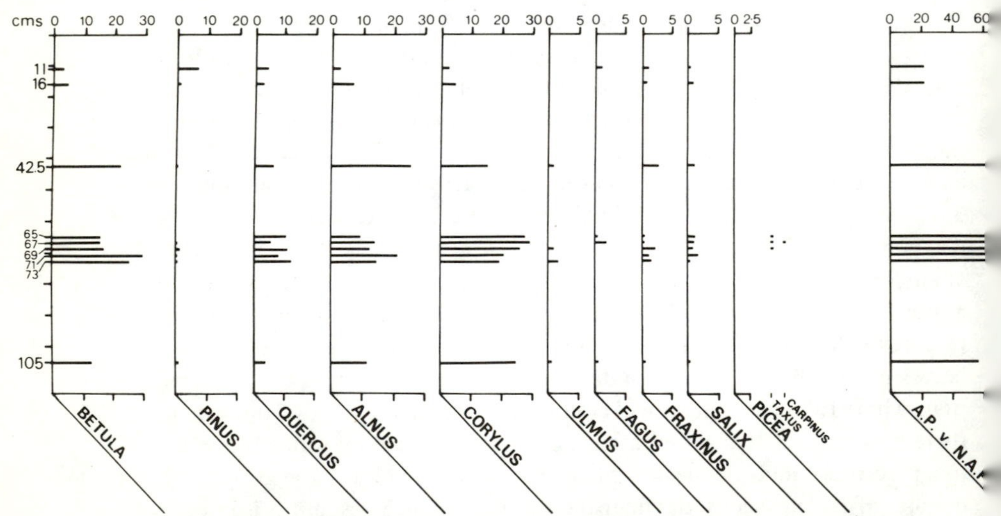

Figure 31. Pollen diagram from monolith HI3. Based on pollen sum of at least 250 non-mire pollen. Summary diagram showing main types only.

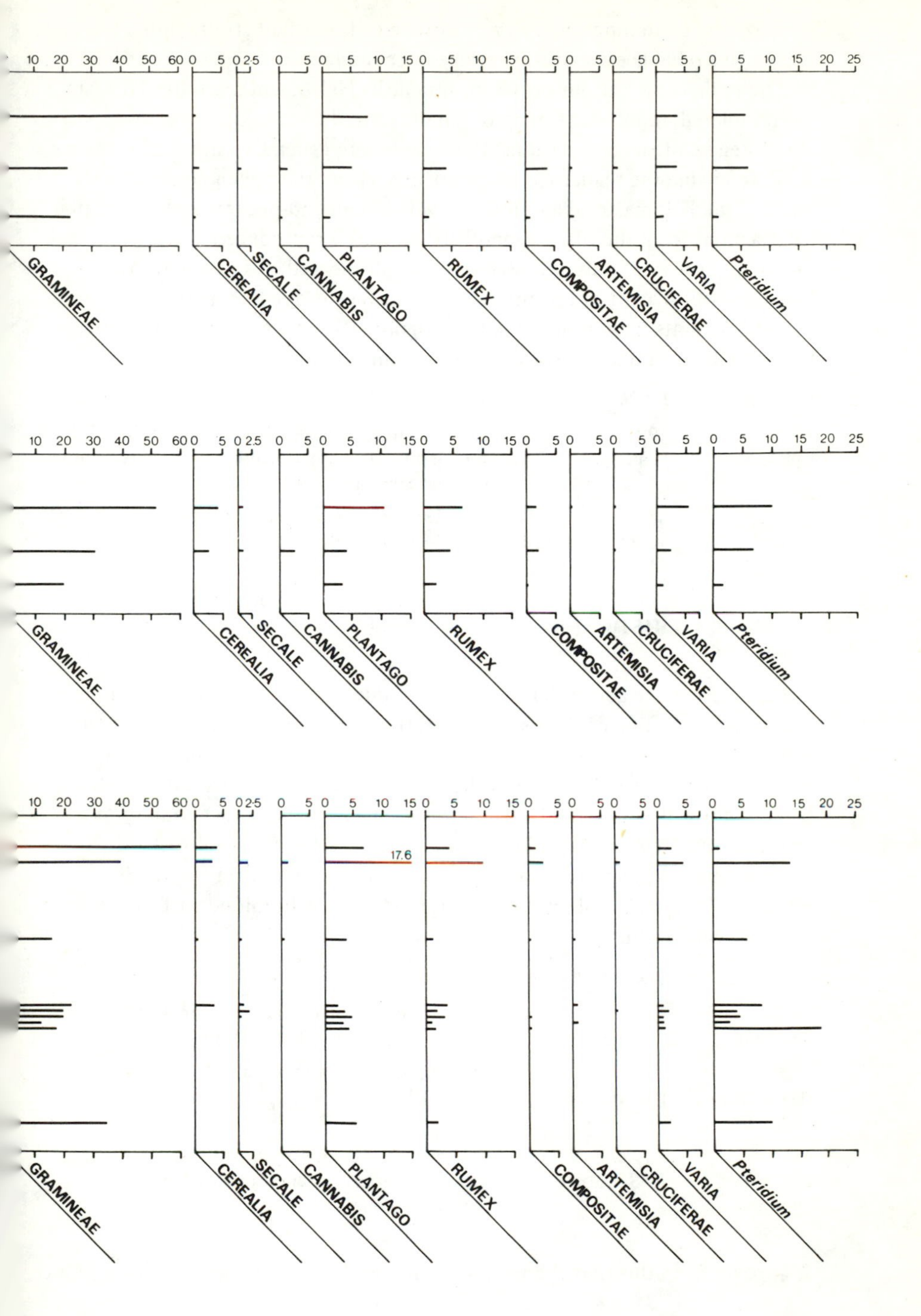
GRAMINEAE
CEREALIA
SECALE
CANNABIS
PLANTAGO
RUMEX
COMPOSITAE
ARTEMISIA
CRUCIFERAE
VARIA
Pteridium
17.6

to prove the contemporaneity or otherwise of important stratigraphic features, and to give reliable estimates of rates of accumulation.

The pollen correlations between monoliths HI1-8, and monolith HI9, are given in detail, in tabular form, to demonstrate the technique employed and the closeness of the correlations; these tables are generally omitted for sections A-G. To enable the reader to independently check the pollen correlations a print of the HI9 master diagram is provided in an end-pocket, and all the pollen diagrams from the other monoliths have been reproduced at the same scale. These are, however, purely illustrative and all the author's correlations are based on 'full-size' diagrams printed on translucent dye-line material.

Pollen counts from monoliths HI1 (figure 29) show an excellent all round fit with HI9 as noted below (all depths in cm):

Depth HI1	*Depth HI9*	*Remarks*
15 Upper pool	19.5 Upper pool	Almost 100 % correspondence – especially AP/NAP and Gramineae. Only exception is Cerealia – more variable generally (= Zone H-I boundary)
38 *S. cusp.* mud of middle pool	43 Middle pool mud	Excellent fit – *Betula* and *Corylus* slightly out; AP, Gramineae and others all correspond (= early Zone G)
50 H4 *Sphagnum* hollow peat	51 H4S lawn peat	Good all round fit – no spectrum more than 5 % out (= Zone F)

It is clear from the above that the main middle and upper pools are contemporaneous at these two ends of section HI; the depth difference of 5 cm is not very significant when one considers the variations of the bog surface and the slight upward curve in the monolith 1 pool – the sample is from the pool edge. If the slight variation is 'real' then it would mean that peat accumulation between 50/51 cm and pool level was slightly greater in monolith 1.

From monolith HI2 only three pollen samples were taken (figure 30) to correlate the upper pools and the overgrowth of the humified surface, with the following results:

Depth HI2	*Depth HI9*	*Remarks*
18.5 Pool mud	19 Pool mud	Exact fit of all curves and indicator species – no *Cannabis* (= Zone H-I boundary)
33.5 H3S above pool mud	43 Pool mud	All good fits except *Quercus* (5 % v. 10 %) (= early Zone G)
45 H3S above humified surface	49 H4S peat	All good fits except *Corylus.* AP and Gramineae; excellent correspondence; cereals and *Cannabis* absent (= Zone F)

It appears from this that the peat in monolith 2, possibly like that from monolith 1, was accumulating slightly faster than that in monolith 9. The pool mud of monolith 2 had given way to an overgrowth of fresh gingery *Sphagnum* while the monolith 9 pool was still in existence and monolith 2 accumulated some 11.5 cm of peat (45 to 33.5 cm) in the same time as monolith 9 accumulated only 6 cm (49-43 cm). This could be connected with differences in *Sphagnum* species (section 5.4), monolith 9 containing abundant *S. cuspidatum* at this level, a species which tends to form felted mats of peat rather than the 'bulked-up' masses of cymbifolian species.

Monolith HI3 has rather more pollen spectra than monoliths 1 and 2. A series of five spectra were taken to fix the date of the lower pool, and further counts done to date the initiation of the upper hummock and its overgrowth, and at 105 cm to 'anchor' the base of the monolith. The results are shown on the pollen diagram, figure 31, and the correlations in the following table:

Depth HI3	*Depth HI9*	*Remarks*
11) 16) Pool mud + *Sphagnum* infill	(15 (19 Pool mud + *Sphagnum* fill	Good fit of AP + Gramineae, and presence/absence of *Cannabis* (Zone H/I boundary, and Zone I).
42.5 Small embryo hummock	45 H4S below middle pool	Good fit – especially AP + Gramineae and *Betula* (= Zone F/G border)
65 67 69 71 73 Series through lower pool (67-71 cm)	65 | 73 Lower pool	Run of five spectra with excellent fit of ± all curves, including *Pteridium* peak; *Plantago, Betula* AP and Gramineae especially good fits (= uppermost Zone D).
105 Humified basal peat	105 Humified basal peat	All curves good fit – agrees in absence of cereals, etc. also (= mid-Zone C).

This series of close correlations demonstrates that in peat section HI hummock and hollow peat grew up more or less in pace with each other. The growth of humified peat in HI3, between 65 and 42.5 cm depth more or less equalled that of unhumified hollow peat in HI9 so that at about 1400 AD the surface at HI3 was a small knot of *Eriophorum-Calluna,* surrounded by *S. cuspidatum* rich wet lawn, just prior to pool formation, the same stage as in HI9. Perhaps rather more surprising however, is the contemporaneity of the upper pools, from which it follows that the upper hummock in HI3 accumulated peat at about the same pace as the wet lawn in HI9.

To further tie down the chronology of the growth of this 'HI3 hummock' a monolith, designated HI3a, was taken across the hummock extension – see

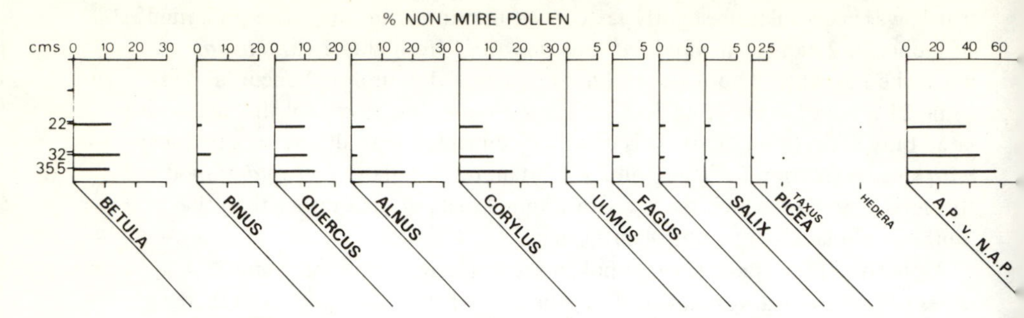

Figure 32. Pollen diagram from monolith HI3a. Based on pollen sum of at least 250 non-mire pollen. Summary diagram showing main types only.

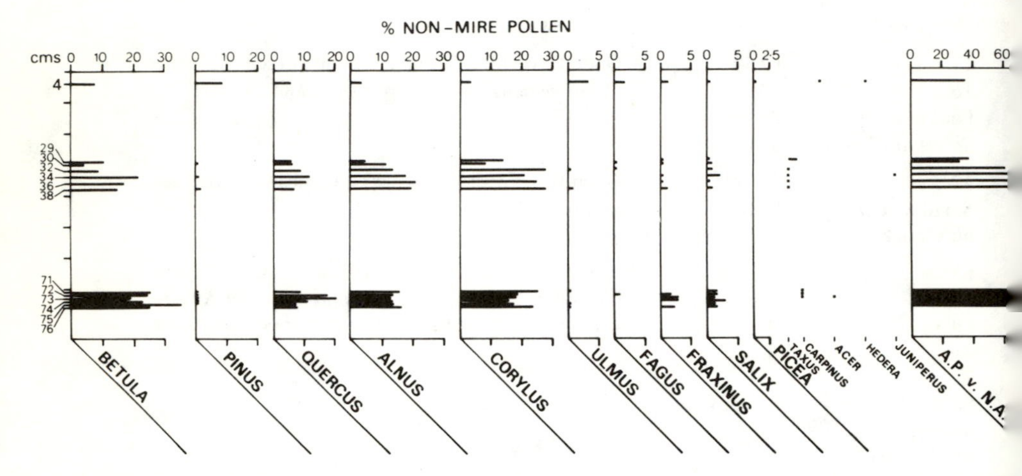

Figure 33. Pollen diagram from monolith HI4. Based on pollen sum of at least 250 non-mire pollen. Summary diagram showing main types only.

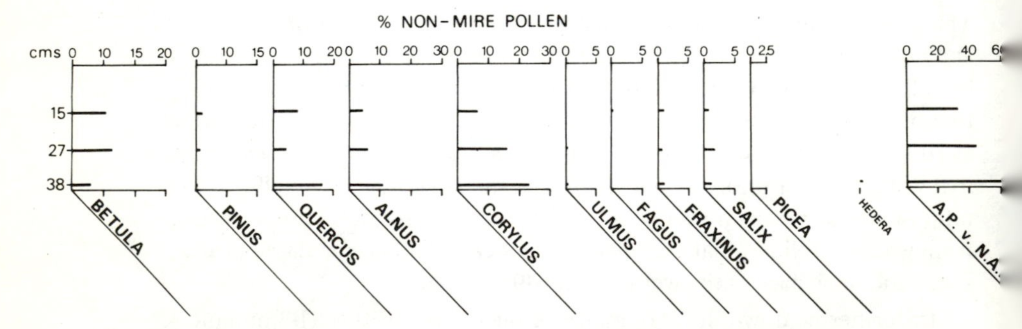

Figure 34. Pollen diagram from monolith HI5a. Based on pollen sum of at least 250 non-mire pollen. Summary diagram showing main types only.

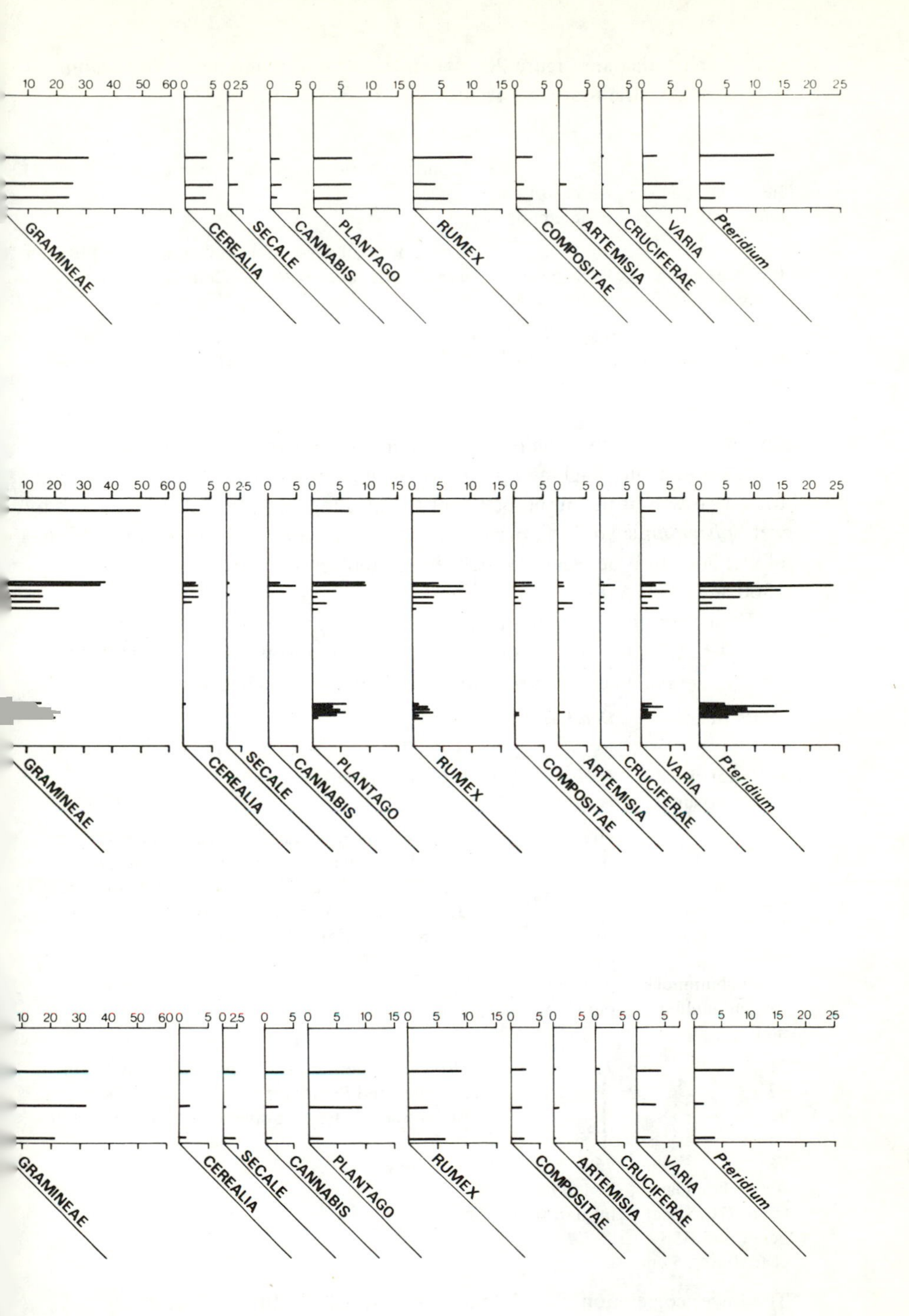
GRAMINEAE
CEREALIA
SECALE
CANNABIS
PLANTAGO
RUMEX
COMPOSITAE
ARTEMISIA
CRUCIFERAE
VARIA
Pteridium

stratigraphic diagram, figure 20. Pollen counts above and below the hummock extension gave the following results, also shown on figure 32.

Depth HI3a	*Depth HI9*	*Remarks*
22 H3S overgrowing hummock	25 H3S overlying middle pool	Good fit or more or less all spectra (= mid Zone H)
32 H5S peat below hummock extension	38 H4S above middle pool	Good fit of all curves except Gramineae – but combined with count below (= Zone G)
35.5 Middle pool mud	42.5 Middle pool mud	Excellent match of all spectra (= lower Zone G)

This match shows that the middle pool muds, though at different depths, are contemporaneous, backing up the evidence from monoliths 1 and 2; and that the outgrowth of the hummock over the infilling pool, then its overgrowth by wet *Sphagnum* lawn, took place at a rate comparable with the accumulation of wet lawn peat between the two upper pools in monolith 9 (between circa 1500 and 1700 AD).

The pollen counts and radiocarbon dates from monolith 4 confirm this general picture, but the radiocarbon assays were not without their problems. Firstly, the pollen correlations showed the following situation:

Depth HI4	*Depth HI9*	*Remarks*
4 H2S peat-check for disturbance	7 *Sphagnum* lawn	A fair fit – merely to check for overturned peat (= Zone I/J border)
29 30 32 34 36 38 Across hummock top and middle pool	40 | 43 | | 49 Across middle pool	Excellent fit of all curves to within a few percent and all in sequence – e.g. start of *Cannabis* and cereals coincide, also Gramineae rise. Pool mud at 32 cm exact contemporary with HI9 pool (= Zone F across into Zone G)
71 72 73 74 76 Across boundary from H7C/S peat below and H5S peat above 75 cm	71 | | | 76 Wet *Sphagnum* infill of pool	Using constraint of low Gramineae, high AP, high *Plantago* and *Pteridium* HI4 71-76 must be in mid/upper zone D. Best fit obtained at 71-76 cm HI9.

The lower correlation, 71-76 cm, is well founded both on pollen-analytical grounds (diagram, figure 33) and in being the first peat of less than humification 6 in both monoliths. However, the pollen correlation gives a date (from the HI9 Age/Depth profile, figure 28) of 890-950 AD – quite at variance with the radiocarbon date (Hv 3079) of 475 ± 85 AD, which is at the zone C/D pollen boundary (figure 27). There is no way in which the pollen from HI4, 71-76 cm, can be matched up with the master diagram at any depth greater than 85 cm (= 780 ± 50 AD) because of the great difference in, especially, the Gramineae percentages (maximum 20 % in HI4, a level not reached again in HI9 until a depth of 112 cm, approximately 160 BC) and when all the spectra are considered together a direct depth-for-depth correlation results, giving the median date of 920 AD. The difference of at least 360 years between the radiocarbon date (taking the 'upper' limit of the one standard deviation) and the 'pollen-correlative' date cannot be brushed aside – although, of course, there is only a 68 % probability that the true age lies between 390 AD and 560 AD – and the possible reasons for this discrepancy is considered later in this section when the two other aberrant dates, from monoliths HI4 (31-34.5 cm) and HI6 (69-72.5 cm), have been reported.

Above this lower wet layer grew a broad hummock of *Sphagnum/Calluna/Eriophorum* culminating in a dry surface at 37 cm, above which grew some 4 cm of fresh *Sphagnum* peat leading to a pool mud at 32-33 cm. The pollen correlation shows that the two middle pool muds are contemporaneous and, furthermore, that there is no hiatus across the humified/unhumified boundary – or at least only a very short cessation of peat growth – for the pollen correlation of 38 cm to 49 cm is followed by 36 = 47, 34 = 45 etc. These correlations give an accumulation rate of 1 cm every 10 years (based on accepting the 1270 AD C^{14} date for 38 cm and the 920 AD 'pollen' date for 73.5 cm; 350 years divided by 35.5 cm equals 1 cm every 9.85 years). The 1270 AD date on the dry hummock top fits in well with the known dryness of the summers in the late 1200's, as well as correlating at exactly the correct level on the HI9 Age/Depth curve. This fairly rapid rate of peat accumulation is not too unusual in *Sphagnum/Calluna/Eriophorum* hummock peats (Walker 1970).

The pool in monolith 4 is shown by pollen correlation to be exactly contemporaneous with that of monolith 9 and the radiocarbon date of 860 ± 135 AD for the HI4 pool, is obviously incorrect on these grounds, as well as being some 400 years older than the date of 1270 AD on the peat 6 cm lower down the profile. Accepting the pollen correlation of the pools and the date of 1445 AD for both of them, the accumulation rate for the peat between 1270 AD and 1445 AD is then 1 cm in 25 years, indicating either a slight hiatus or compression of the lax growing *S. cuspidatum,* or that the pool mud represents several years of open water accumulation. Naturally, a combination of all three factors is quite likely.

Pollen counts were performed on monolith 5a (figure 34) in an attempt to date the hummock outgrowth, with the following results:

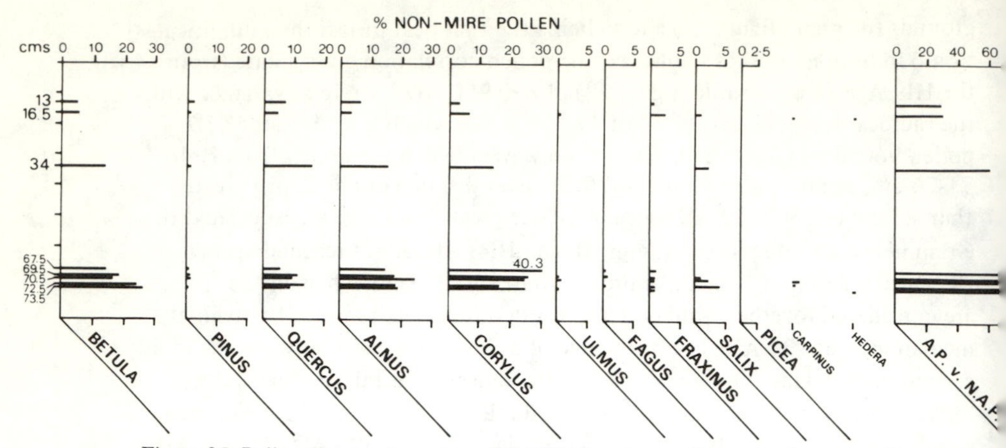

Figure 35. Pollen diagram from monolith HI6. Based on pollen sum of at least 250 non-mire pollen. Summary diagram showing main types only.

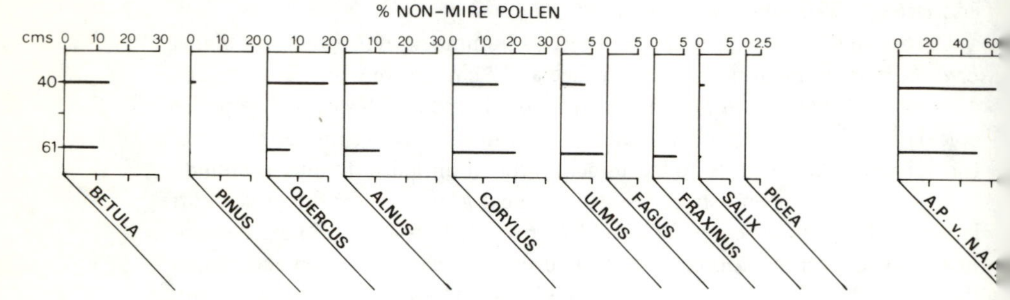

Figure 36. Pollen diagram from monolith HI7. Based on pollen sum of at least 250 non-mire pollen. Summary diagram showing main types only.

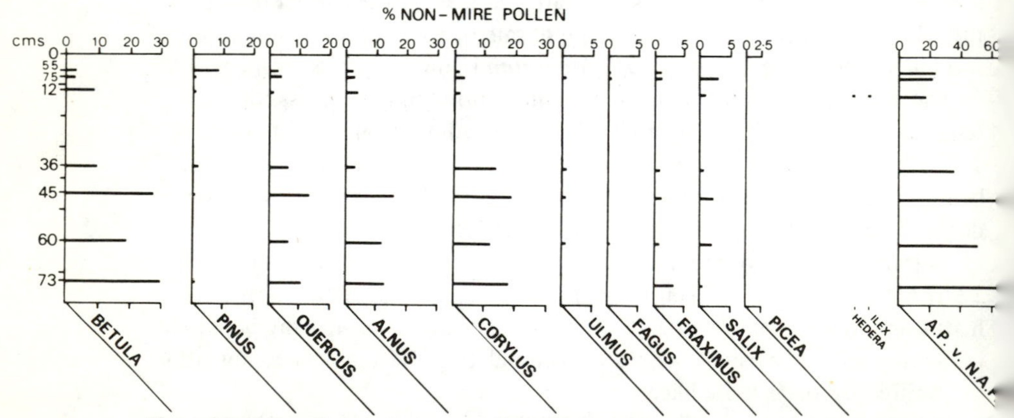

Figure 37. Pollen diagram from monolith HI8. Based on pollen sum of at least 250 non-mire pollen. Summary diagram showing main types only.

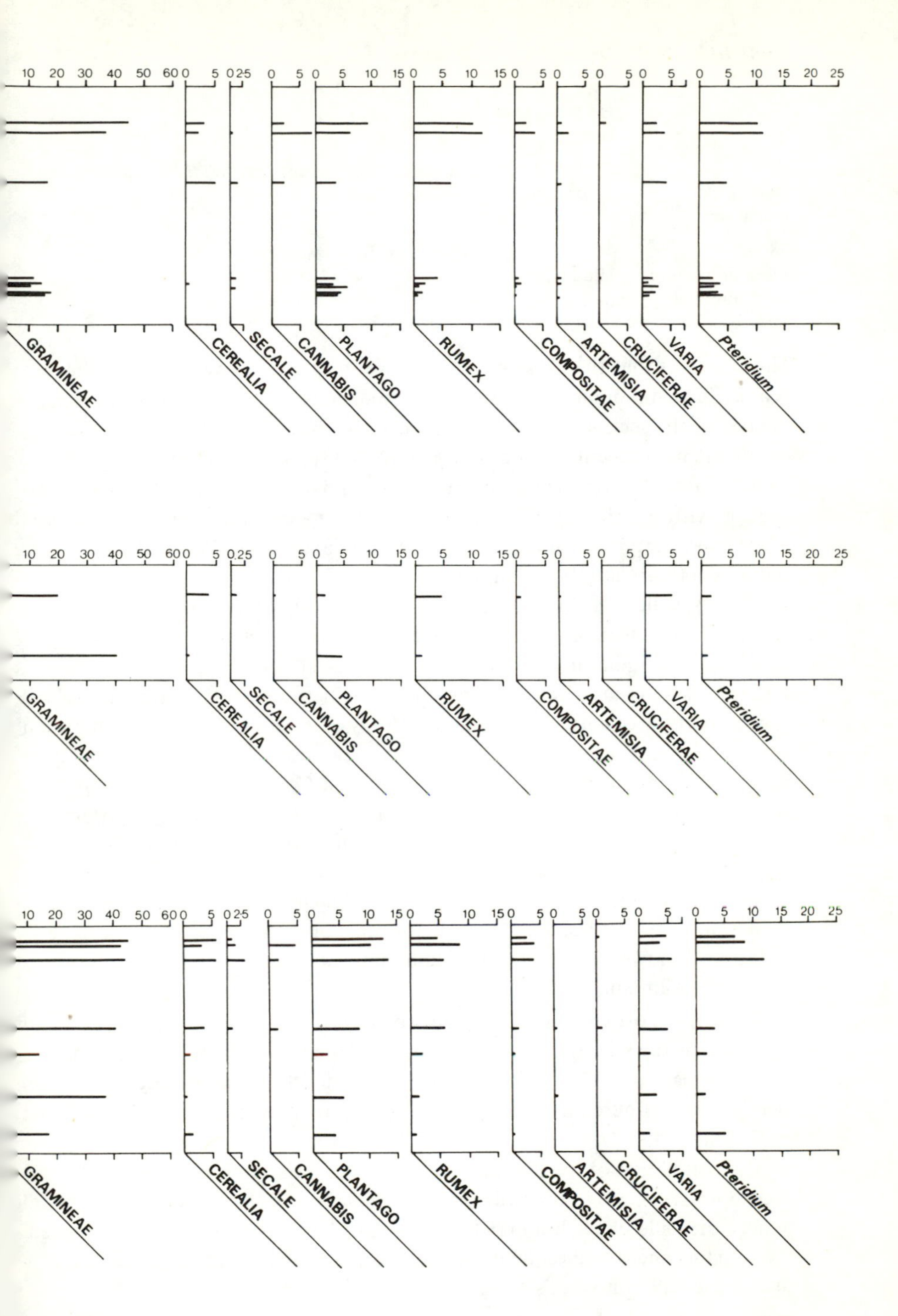
10 20 30 40 50 60
GRAMINEAE
CEREALIA
SECALE
CANNABIS
PLANTAGO
RUMEX
COMPOSITAE
ARTEMISIA
CRUCIFERAE
VARIA
Pteridium
10 20 30 40 50 60
GRAMINEAE
CEREALIA
SECALE
CANNABIS
PLANTAGO
RUMEX
COMPOSITAE
ARTEMISIA
CRUCIFERAE
VARIA
Pteridium
10 20 30 40 50 60
GRAMINEAE
CEREALIA
SECALE
CANNABIS
PLANTAGO
RUMEX
COMPOSITAE
ARTEMISIA
CRUCIFERAE
VARIA
Pteridium

Depth HI5a	*Depth HI9*	*Remarks*
15 H3 *Sphagnum* lawn	24 H3 *Sphagnum* lawn	A good fit of all curves (= mid-Zone H)
27 Hummock extension	27 *Sphagnum* lawn	Quite good fit, except for *Pteridium* and *Corylus* (= lower Zone H)
38 Fresh *Sphagnum* over humified surface	43 Middle pool mud	Excellent fit (= lower Zone G)

This indicates that the phase of renewed wet peat growth over the humified peat surface in monolith 5a was contemporaneous with pool formation over the rest of the section, and thereafter peat accumulation was slower in the non-pool environment of monolith 5a with the spread of hummock species from the right to give the hummock extension dated at circa 1670 AD by correlation with the HI9 Age/Depth curve. It then appears that there was a phase of extremely rapid accumulation in monolith 5a; between 27 cm (1670 AD) and 15 cm (correlates best with HI9 24 cm = 1715 AD) the rate is 1 cm in 3.75 years while it is about 1 cm in 15 years in HI9. While the former rate is high it is not abnormally so; both Turner (1964) and Schneekloth (1965) have reported similar rates from unhumified raised bog peat (3.3 years/cm and 4.5 years/cm respectively). The difference between the two monoliths may be partly explained by differential decay and compaction factors between species. The peat overgrowing the hummock extension is primarily of highly elastic cymbifolian species with a fair amount of *Calluna* twigs which may act as 'scaffolding' for the Sphagna, and the differential decay factor, according to Clymo (1965), can produce rates of decay in *Sphagnum papillosum* which are only half that of *S. cuspidatum.*

Monolith HI6 is an important one, showing a clear progression from pool to hummock to wet hollow. The pollen counts on HI6 (figure 35) were aimed at dating the lowermost pool, the mid-hummock area and the time of hummock-top swamping.

The uppermost correlation is not surprising, though it is interesting in demonstrating that the swamping of the hummock top at 16.5 cm took place prior to the formation of the open water pool in monolith 9 – according to the Age/Depth profile (figure 28), the hummock overgrowth dates from 1740 AD, the pool from 1785 AD.

The correlation of the 34 cm level in the HI6 hummock with the middle pool at 43 cm in monolith 9 (dated to 1425 AD), and therefore with the middle pools either side of the hummock, shows that the hummock only just survived the flooding and its subsequent upward growth was at about the same rate as that in the HI9 hollow.

Depth HI6	*Depth HI9*	*Remarks*
13 16.5 H3-4 wet S peat overgrowing hummock top	19 22 Transitional peat leading to pool mud	Generally good fit; *Rumex* and *Betula* only exceptions (= upper Zone H)
34 H6 peat-centre of hummock	43 Middle pool mud	Excellent fit all round (= lower Zone G)
67.5 69.5 70.5 72.5 73.5 *S. cusp.* pool	66.5 | 70 | 72 Dry to wet S lawn	Good fit of almost all curves, at individual levels and in sequence (= upper Zone D)

The lowermost pool in monolith 6 was dated by radiocarbon assay to 1360 ± 65 years BP, 590 AD (Hv 3080). The pollen correlation dates the beginning of the pool much later than this, to around 950 AD, and at almost the same depth value as in monolith 9. The pollen spectrum at a depth equivalent of 590 AD would have some 15 % more Gramineae pollen than these HI6 samples and there are various other incompatabilities – the radiocarbon date must therefore be incorrect.

In all, three radiocarbon dates are judged to be 'too old' on the grounds described below:

Lab. No.	*C^{14} date*	*Po date*	*Yrs 'too old'*	*Peat type*
Hv3077	860 ± 135	1445	585	H3S peat + pool mud
Hv3079	475 ± 85	920	445	H5S/C wet 'regrowth' peat
Hv3080	590 ± 65	950	360	H6 *S. cuspidatum* peat above humified surface

Above dates are 'AD'.

It will be noted that all three dates are from wet lawn or pool peat representing a regrowth on top of an older drier surface. The 'correct' C^{14} date of pool-mud peat in monolith HI9 was not from such a 'regrowth situation' – the peat above and below the pool-mud was only of humification 4. This fact may point to the reason why the dates are incorrect, and wrong in the opposite way to that usually encountered in radiocarbon work, by being 'too old' rather than 'too young'! The latter phenomenon is quite plausibly explained by assuming that the sample has been penetrated by rootlets from above, or that a small amount of highly active 'bomb carbon' has contaminated the sample.

In the case of these three samples of pool peat 'old carbon' may have been incorporated into the sample by either mobilisation of carbon of the old surface due to metabolisation by renewed plant growth, or by erosion of the old,

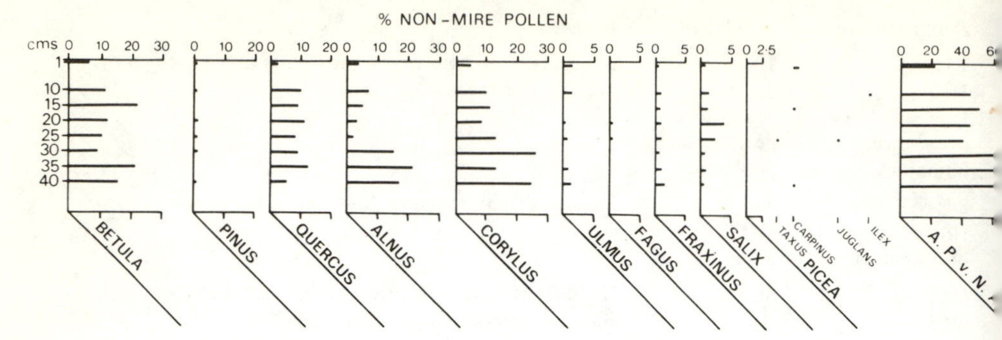

Figure 38. Pollen diagram from monolith AI1. Based on pollen sum of at least 250 non-mire pollen. Summary diagram showing main types only.

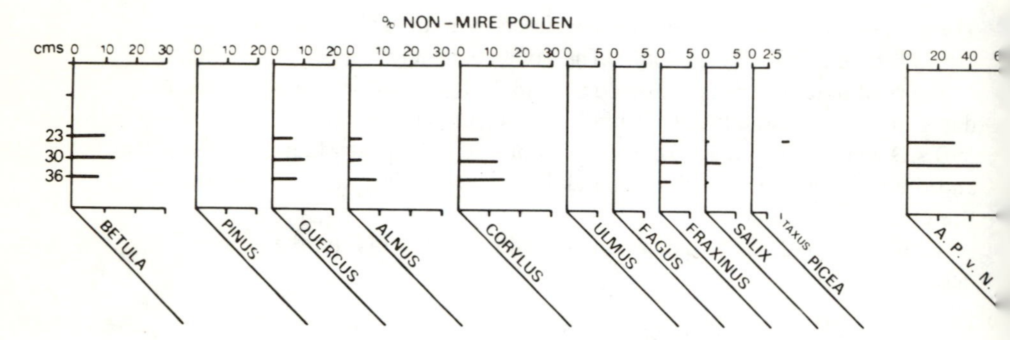

Figure 39. Pollen diagram from monolith AII1. Based on pollen sum of at least 250 non-mire pollen. Summary diagram showing main types only.

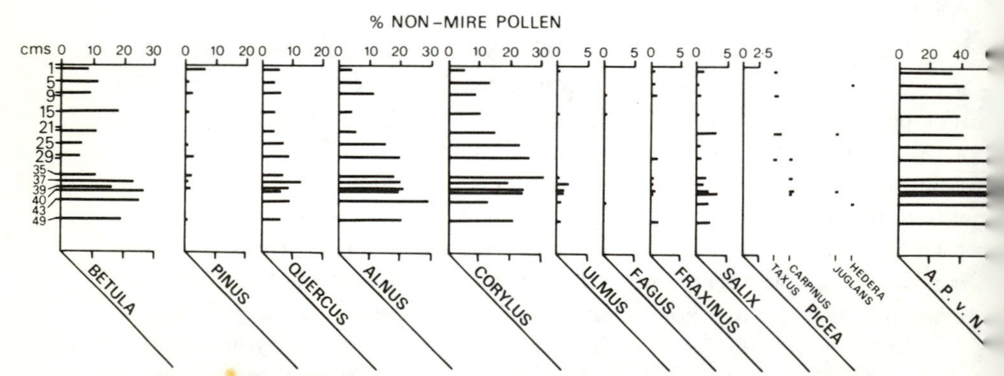

Figure 40. Pollen diagram from monolith AI2. Based on pollen sum of at least 250 non-mire pollen. Summary diagram showing main types only.

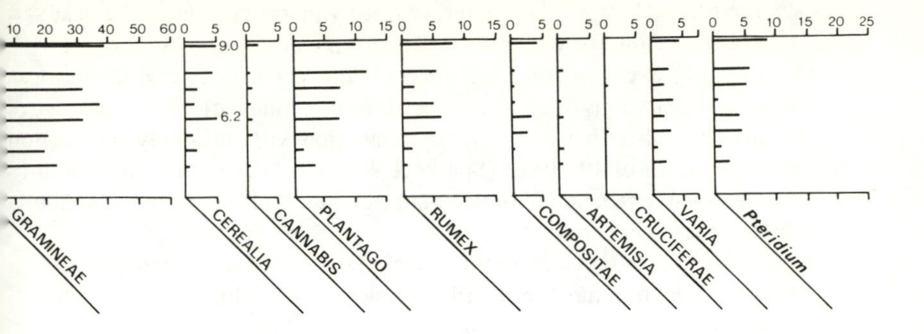
10 20 30 40 50 60
0 5
9.0
6.2
0 5
0 5 10 15
0 5 10 15
0 5
0 5
0 5
0 5
0 5 10 15 20 25
GRAMINEAE
CEREALIA
CANNABIS
PLANTAGO
RUMEX
COMPOSITAE
ARTEMISIA
CRUCIFERAE
VARIA
Pteridium

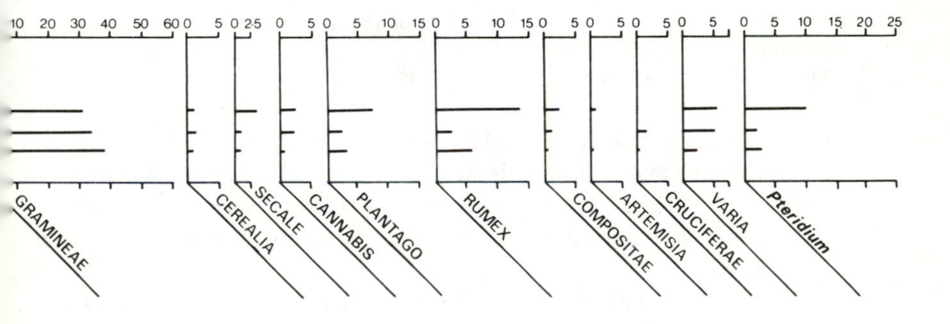
10 20 30 40 50 60
0 5
0 2.5
0 5
0 5 10 15
0 5 10 15
0 5
0 5
0 5
0 5
0 5 10 15 20 25
GRAMINEAE
CEREALIA
SECALE
CANNABIS
PLANTAGO
RUMEX
COMPOSITAE
ARTEMISIA
CRUCIFERAE
VARIA
Pteridium

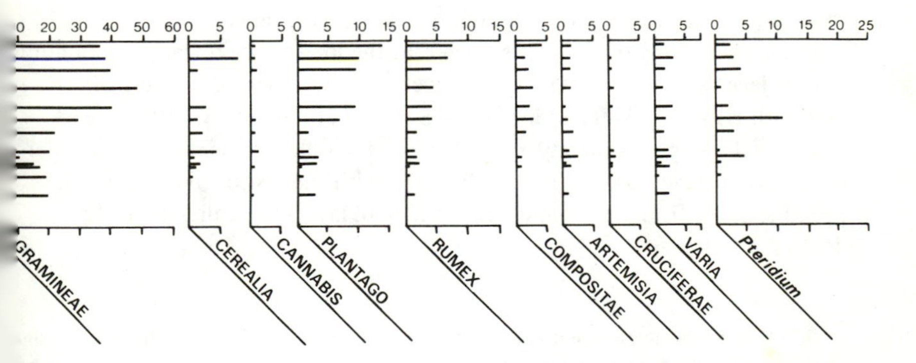
0 20 30 40 50 60
0 5
0 5
0 5 10 15
0 5 10 15
0 5
0 5
0 5
0 5
0 5 10 15 20 25
GRAMINEAE
CEREALIA
CANNABIS
PLANTAGO
RUMEX
COMPOSITAE
ARTEMISIA
CRUCIFERAE
VARIA
Pteridium

dry surface material by wind or water, or else by incorporation of the dead stems of *Calluna* from the old surface. These three possibilities are not, of course, mutually exclusive; they were accepted by the radiocarbon laboratory as probable explanations (M.A. Geyh, personal communications).* There is the further possibility that only 5 % contamination with 'infinitely old' carbon would give an error of 400 years (Sparks & West 1972), very near the amount of error detected here, and colloidal organic carbon is also known to be mobile within the ecosystem (Lock *et al* 1977).

Monolith HI7 was taken merely to check the dates of the outgrowth and overgrowth of the hummock primarily sampled by monolith 6 as well as to study the plant succession. The results are shown below and on figure 36.

Depth HI7	*Depth HI9*	*Remarks*
40 H3S above hummock	44 Just below middle pool	Very good fit of almost all curves (= lowest Zone G)
61 H3S below hummock	59 H3S above wet layer	Good fit of almost all curves (= mid Zone E)

These correlations give dates of 1125 AD for the lower sample and 1400 AD for the upper so that, in round figures, one can date the hummock outgrowth to between 1200 and 1350 AD, a growth rate of 1 cm every 13 years, its overgrowth being part of a general return to wetter conditions over the whole of section HI.

Monolith 8 was taken in order to study the infill of a hollow between two large hummocks, the uppermost part of the monolith being overgrown by the sideways spread of one of the hummocks.

Pollen correlations (figure 37) between monolith 8 and 9 were performed in order to date the lower and middle pools and the approximate rate of hollow infill, and to date the surface hummock in relation to the adjacent pool in monolith 9. These pollen correlations are tabulated overleaf:

From these correlations it is shown that the lower pool formed over a humified surface at about 965 AD, which agrees very well with the date from monolith HI6 (950 AD) and infilled at a rate of about 1 cm every 15 years (60 cm, HI8 = approximately 1160 AD). The middle pool is shown to be contemporaneous with that in monolith 9, and with the rest of the section, the pollen samples from below and above the pool layers matching depth for depth.

* These discrepancies were discussed at length with Professor Geyh in Hanover in June 1978. No wholly satisfactory explanation seems possible and check-dates can not now be performed since the site has been cut and destroyed.

Depth HI8	*Depth HI9*	*Remarks*
5.5) 7.5) H8 Black hummock peat	19 Upper pool mud	Samples bracket 19 cm level in HI9; excellent fit of all curves, especially *Pinus.* ∴ 6.5 cm HI8 = 19 cm HI9 (= Zone H/I border)
12 H5 *Sphagnum* lawn with Calluna	20 Start of upper pool-mud	Must fit at this level, otherwise AP, Gramineae, etc. far out (= top Zone H)
36) 45) H3S above and below middle pool-mud	(36 (45 H3S above and below middle pool-mud	Excellent fit, and in sequence (= start of, and middle, Zone G)
60) 73) Lower pool and infill	(57 (70 Lower hollow infill	Excellent fit of all spectra at both levels (= upper Zone D and Zone E)

The correlations of the hummock and pool at the top of monoliths 8 and 9 are, however, somewhat surprising. All three samples from monolith 8, at 5.5, 7.5 and 12 cm, match with the pollen spectra of monolith 9 at the 19-20 cm level, though differentiation between the three samples is quite possible based on crucial indicators such as the pine, birch and hemp frequencies (figure 37). This is either to be explained by very rapid hummock growth in monolith 8 or a fairly long-standing pool in monolith 9, or a combination of both factors. The age/depth profile of monolith 9 (figure 28) gives a growth rate of 1 cm every 14.9 years to the zone including the 19-20 cm level, so on this basis the hummock would have grown, between 12 and 6.5 cm, at a rate of a centimetre every 2.7 years, not an impossibly high figure in surface peats (Aaby 1977). On the other hand the pool-mud in monolith 9 is very well-developed and 1.5 cm thick (18.5-20 cm). Estimates of the time required for the build-up of such algal muds are hard to come by, although Boatman & Tomlinson (1973) demonstrate the persistence and enlargement of a particular pool over at least 23 years and Boatman (1977) considers that ombrotrophic bog pools are sub-optimal habitats for *Sphagnum cuspidatum,* the implication being that they may take several years to infill. Naturally this will also be related to the size and depth of the pool, and one cannot imagine the HI9 pool being open for more than a few decades, but if this line of reasoning is accepted and a figure of 30 years taken as the time needed to produce the centimetre of algal mud between 20 and 19 cm (very near the modal values for nekron muds and fine detritus mud given by Walker (1970, pp.128-132)), then the corresponding rate of hummock growth drops to 1 cm in 5.5 years. It is clear from this that hummock-type peats are capable of greater growth rates than is commonly accepted – an important point in the argument over hummock-hollow succession and one to be discussed later.

Monolith 9, as already discussed, consists of hollow peat throughout, with changes from a drier to a wetter regime of peat formation at 775 AD, 1030 AD, 1190 AD, 1425 AD (pool-mud) and 1785 AD (pool-mud).

5.3.3 Sections AI and AII

Three pollen diagrams were constructed from these sections to give dates on the pool formation and infill to the right of the sections and on the growth of the strongly humified hummock to the left (figure 14). The diagram from monolith AI1 (figure 38) correlates best with the master HI9 diagram in two separate sections. The counts from 1-25 cm correlate to within a few percent with HI9, 19-42 cm. This dates AI1, 1 cm, at 1785 AD which is entirely in accord with the stratigraphic evidence of drying-out and rehumification of the uppermost 8 cm, and the documentary evidence of drainage at 1800 AD (figure 12). The counts then all correlate in sequence with 25 cm depth dating to 1445 AD (HI9: 42 cm). The lower three counts form a separate group; 30, 35 and 40 cm AI1 correlate with 44, 49 and 54 cm HI9.

These correlations mean that the change from H8 peat to H4 peat and the formation of a wet *Sphagnum* lawn started around 1200 AD but the pool-mud formed somewhat later, between 1325 AD and 1365 AD, after which there was very rapid infill at first (1 cm in nine years between 1400 AD and 1445 AD, 30-25 cm in monolith AI1 but only 44-42 cm in monolith HI9) which then slowed down to 1 cm every 14 years as the hollow infilled. The macrofossil diagram (figure 60) showing the humification increasing upwards is in line with this.

The pool sampled later from section AII, monolith AII1, is very similar to that described above and only three pollen counts were performed, merely to date the pool-mud (figure 39). The pollen spectrum from this level, 36 cm AII1, correlates with 43 cm in HI9, that is about 1420 AD and more or less contemporaneous with the main middle pool in HI9.

The hummock in monolith AI2 was dated by a 13-spectrum pollen diagram (figure 40). The correlations between this and pollen diagram HI9 were extremely close and agreed best if diagram AI2 was considered in two parts. Counts from 1-29 cm in AI2 agreed very well with those from 19-47 cm, while the lower block of close counts from AI2, 35-49 cm, agreed best with HI9: 65-78 cm. This results in the lower part of the hummock, including the H5 *Sphagnum* band, growing at a rate of 1 cm every 11 years (49 cm = 885 AD, 35 cm = 1040 AD), and then growth slowing down dramatically between 35 and 29 cm (1040 AD – 1335 AD) to a rate of 1 cm every 49 years. Overgrowth or rejuvenation of this hummock at 24 cm depth is then dated to 1420 AD and the growth rate up to the 1800 AD surface increased again to 1 cm every 16 years.

Hummock and pool development are therefore seen to have gone through related changes of humification at about the same time, a result more in accord with an overall hydrological change in the bog rather than an autogenic cycle.

5.3.4 Section B

The pollen diagram from monolith BI1 shows an extremely good correlation with HI9 on a depth for depth basis, even though the two sites are some 1,000 m apart. The diagram (figure 41) thus dates the strongly growing hummock from 50 to 27 cm as being in existence from at least 1270 AD until 1670 AD, when it was overgrown by a swathe of H3 *Sphagnum.* This *Sphagnum* lawn was then superseded by a layer rich in *S. cuspidatum* and greasy with algae; this very wet phase, from 18.5-15 cm, coincides with the upper pools of section HI, dated from around 1800 AD. Thereafter a drier lawn of cymbifolian Sphagna continues to the surface. There is therefore no suggestion of cyclic processes at work in section BI.

5.3.5 Section CI

Three monoliths were taken from this section (figure 16) and a pollen diagram constructed for each. Additionally a radiocarbon date was obtained from monolith CI2.

The diagram from monolith CI1 (figure 42) has only four pollen spectra but correlates well with diagram HI9. The 35 cm count equates with the 19 cm level in HI9, again showing that the pool-mud in CI1 dates from around 1800 AD. The growth of the hummock below this pool is shown by these pollen correlations to have been well underway at 1240 AD (76 cm: CI1 = 52 cm: HI9) and the growth rate up to 64 cm (= 40 cm HI9, 1475 AD) was a modest 19.6 years/cm. Thereafter the hummock appears to have grown more rapidly and to have been overgrown by a mixed *Sphagnum* lawn at 52 cm (growth rate 9.0 years/cm between 64 cm and 43.5 cm, 1475 AD to 1660 AD), after which growth slowed to 16.5 years/cm until pool-formation just before 1800 AD.

The pollen diagram from monolith CI2 (figure 43) was the most detailed record from the bog, prior to the construction of diagram HI9. Ignoring the top 30 cm, which clearly show an artificial reversion due to ditch-spoil dumping on the 1800 AD surface, the diagram follows the main trends of HI9. However, as was encountered by Turner (1975) in her work on Bloak and Kennox Mosses, there are discrepancies between the diagrams, both in the values of birch and alder, trees which could have produced local differences in pollen fall-out in a bog situation and in levels of Gramineae – perhaps due to local clearance differences. Paying particular attention therefore to more regional indicators such as pine and various herbs which could not have grown on the bog (for example, hemp), one may fix the pollen count for 40 cm CI2 as being

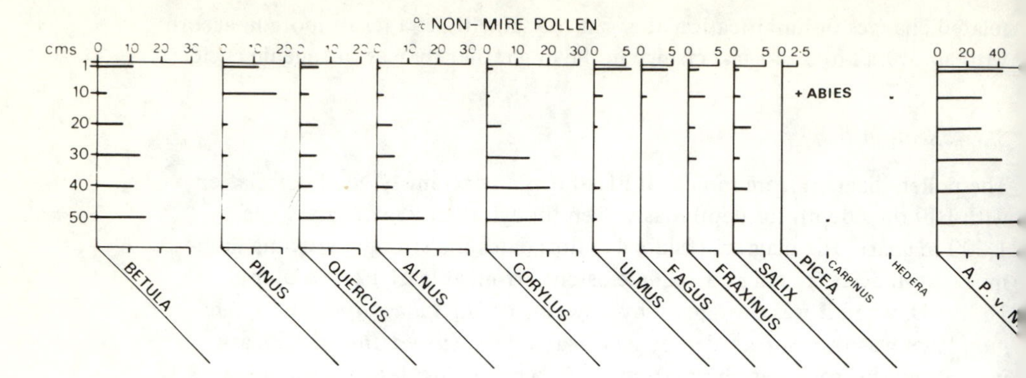

Figure 41. Pollen diagram from monolith BI1. Based on pollen sum of at least 250 non-mire pollen. Summary diagram showing main types only.

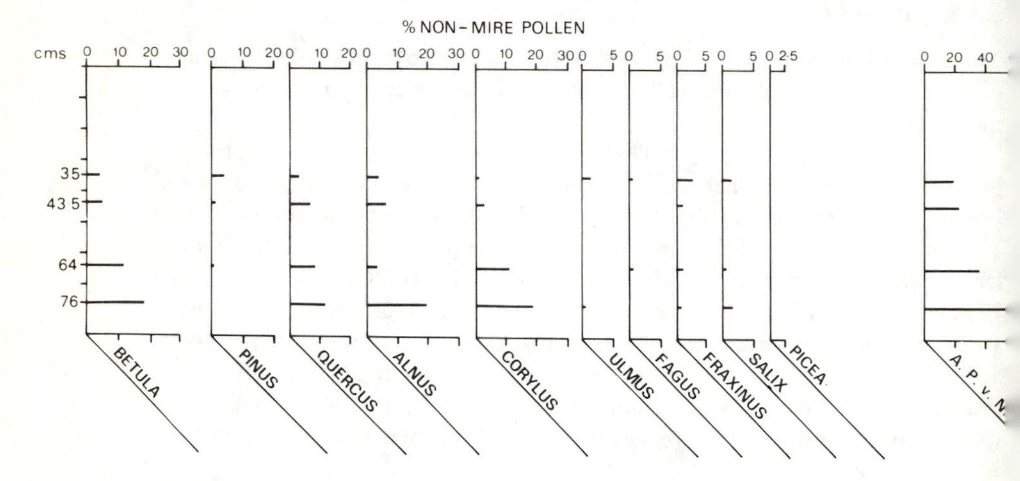

Figure 42. Pollen diagram from monolith CI1. Based on pollen sum of at least 250 non-mire pollen. Summary diagram showing main types only.

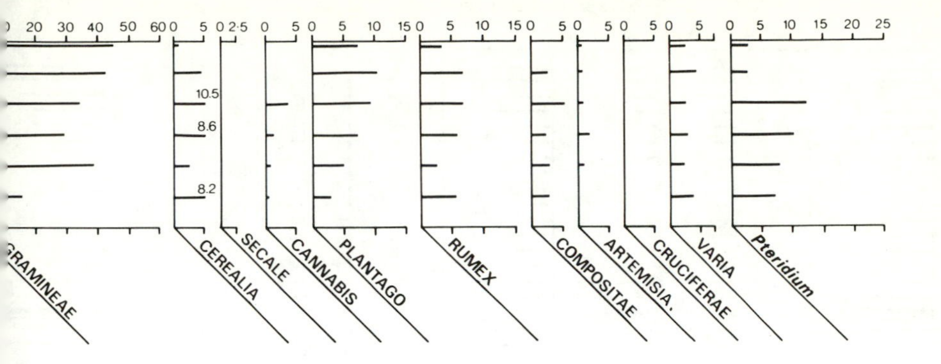
20 30 40 50 60
0 5
0 2·5
0 5
0 5 10 15
0 5 10 15
0 5
0 5
0 5
0 5
0 5 10 15 20 25
10.5
8.6
8.2
GRAMINEAE
CEREALIA
SECALE
CANNABIS
PLANTAGO
RUMEX
COMPOSITAE
ARTEMISIA.
CRUCIFERAE
VARIA
Pteridium

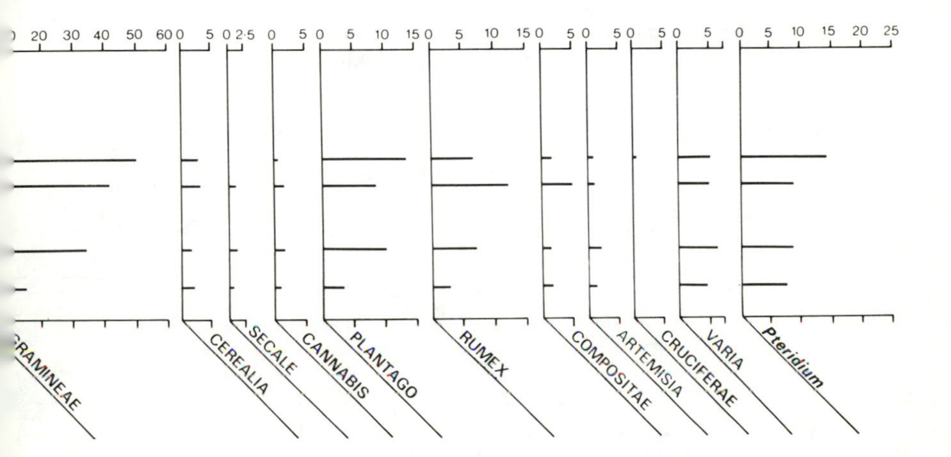
20 30 40 50 60
0 5
0 2·5
0 5
0 5 10 15
0 5 10 15
0 5
0 5
0 5
0 5
0 5 10 15 20 25
GRAMINEAE
CEREALIA
SECALE
CANNABIS
PLANTAGO
RUMEX
COMPOSITAE
ARTEMISIA
CRUCIFERAE
VARIA
Pteridium

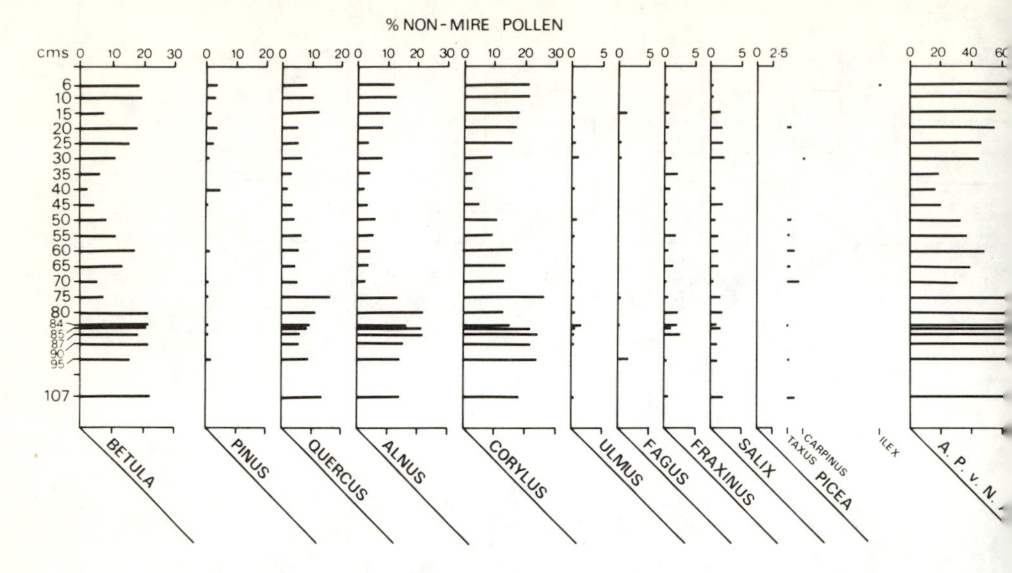

Figure 43. Pollen diagram from monolith CI2. Based on pollen sum of at least 250 non-mire pollen. Summary diagram showing main types only.

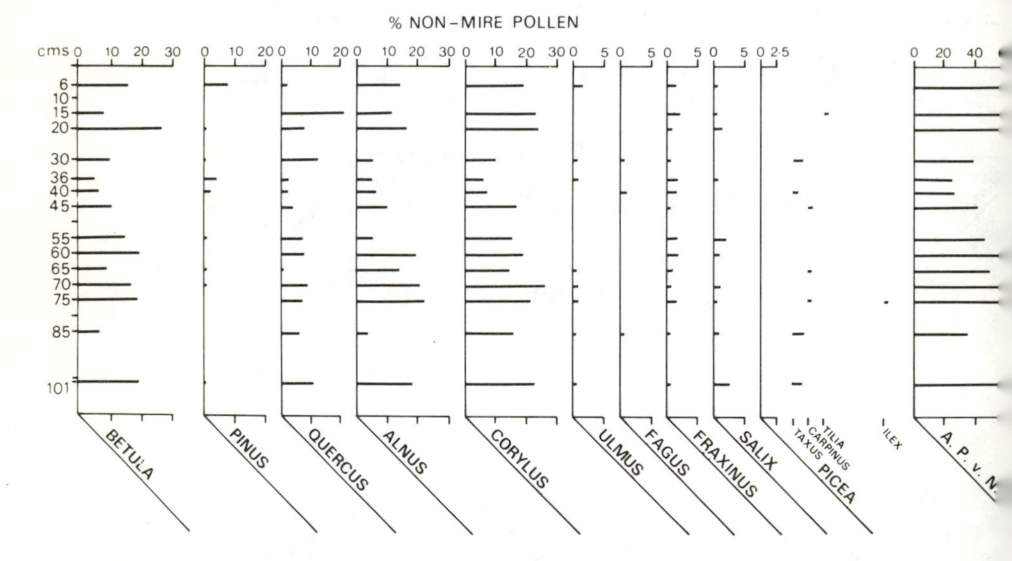

Figure 44. Pollen diagram from monolith CI3. Based on pollen sum of at least 250 non-mire pollen. Summary diagram showing main types only.

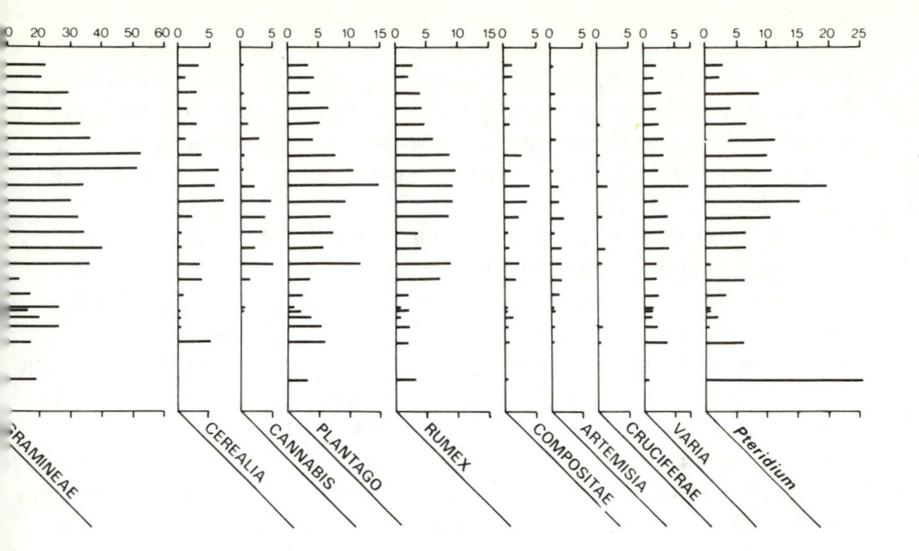
GRAMINEAE
CEREALIA
CANNABIS
PLANTAGO
RUMEX
COMPOSITAE
ARTEMISIA
CRUCIFERAE
VARIA
Pteridium

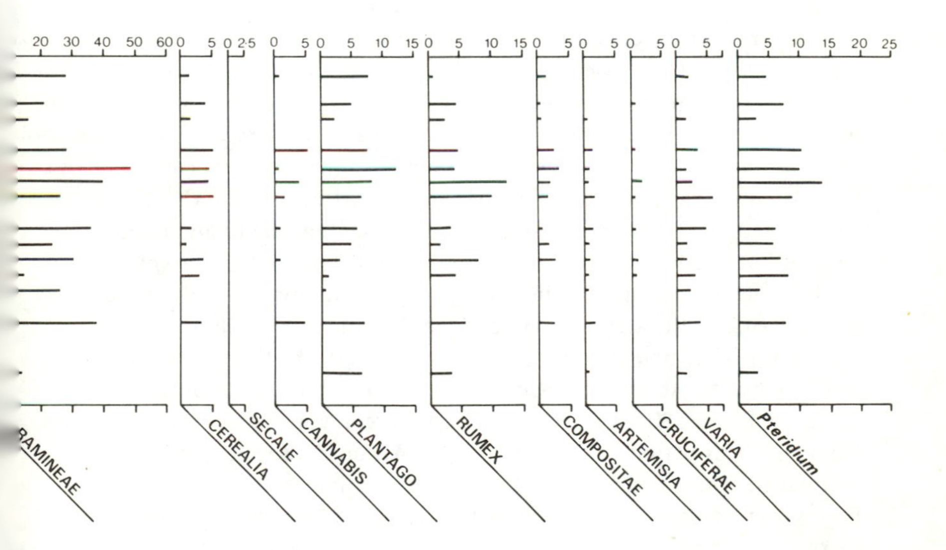
GRAMINEAE
CEREALIA
SECALE
CANNABIS
PLANTAGO
RUMEX
COMPOSITAE
ARTEMISIA
CRUCIFERAE
VARIA
Pteridium

equal to 19 cm HI9, that is 1785 AD. If the radiocarbon date of 940 ± 120 AD (from the Gakushuin University laboratory, Tokyo) is accepted as the true date of the 81-85 cm interval, say 83 cm, then the peat accumulation rate between the two levels is 19.6 years/cm, very similar to many other monoliths. Numerous attempts at matching the pollen curves at different levels gave accumulation rates between 16 and 22 years/cm, so the 19.6 years/cm rate is acceptable from pollen correlative point of view.

At this accumulation rate the first pool mud of 940 AD (the same period as the pools of monoliths HI4 and HI6) was preceded by a humification change, from H6 to H4, at 625 AD, and a lowermost pool-mud at around 700 AD. Above the 940 AD pool there is another, less well-marked, mud at around 1160 AD after which the hollow was encroached upon by a hummock growing from the left, the change to H7 hummock peat being dated to 1550 AD. This hummock formation persisted until buried in 1800 AD.

Pollen diagram CI3 (figure 44) is a much better fit with HI9; it is not as good a fit with CI2 as might be expected from monoliths only 30 cm apart but the discrepancies are not great and could perhaps be attributed, at the 80-85 cm depth for example, to reflotation of pollen grains, CI2 being from the middle of the pool (at 83 cm), CI3 being from the very edge of the same pool (at 81 cm depth).

According to this pollen correlation the start of sedimentation in CI3, at 101 cm, dates to 845 AD and between this and 85 cm (1065 AD) the growth rate was 13.8 years/cm. The lower pool, starting at 82.5 cm appears then to be in a period of faster accumulation (85-75 cm correlates with 1065-1145 AD on the HI9 scale) and dates from 1085 AD. This is at variance with the C^{14} date for the centre of the same pool (940 ± 120 AD), but, as noted above, the pollen spectra do not allow a close correlation and the very edge of such a pool, abutting onto a *Calluna-Eriophorum* hummock, might be expected to be of a later date than the pool centre. If the younger extreme of the C^{14} date range is taken (1060 AD) the discrepancy is, of course, reduced to an insignificant amount but it is thought best to rely on the pollen correlation in monolith 3, the C^{14} date in monolith 2. The discrepancy of a hundred or so years is in any case not vital to the argument concerning the hummock-hollow succession.

Following the pool phase then, at around 1000 AD, there followed a period of rapid infill, at about 8 years/cm up to 75 cm from whence until the burial of the profile in 1800 AD the pollen curves are in good agreement with those from HI9. This period, 75-36 cm, or 1145 AD until 1800 AD, gives a mean accumulation rate of 16.8 years/cm in primarily hummock peat. This formation is overgrown by fresher hollow peat at 40 cm, the junction being notably sharp. This dates the change to 1720 AD, with a pool-mud formation beginning in 1770 AD and only just filling with a little fresh *Sphagnum* before being buried. The synchroneity and transgressive nature of this upper pool is at odds with a cyclic alternation of hummocks and pools, and even though the lower pool and the structures above it appear to give some support to the Osvald view the above datings show it to be contemporaneous with a widespread formation of pools.

5.3.6 Section DI

The pollen counts from this profile were spaced so as to date two episodes of change from humified peat to unhumified peat, at 56 and 16 cm (figure 17). The pollen diagram (figure 45) is in good agreement with that from HI9, with only one or two spectra a few percentage points apart.

Four spectra across the major humification boundary, 65-50 cm: DI1, correlate with 58-43 cm: HI9. This covers the period 1145 AD to 1425 AD, the humification change at 56 cm taking place around 1315 AD. There is an increase in the mean accumulation rate from 18.6 years/cm (65-50 cm) to 10.3 years/cm between 50 and 15 cm – the latter level being correlated with 1785 AD on the basis of three pollen counts at 14, 16 and 18 cm. With this mean rate the pool-mud at 48-47 cm is dated to 1445-1455 AD – exactly contemporaneous with the HI9 main pool. Thereafter the pool infill includes a drier phase, between 1510 and 1610 AD, within a generally wet hollow and then drier communities spread over the hollow around 1705 AD (23 cm), lasting until 1775 AD (16 cm) when a wet *Sphagnum* lawn overgrew the dry surface. The almost exact correspondence of these shifts to a wetter surface in both section D and section H is striking and again argues against local, small cyclic changes.

5.3.7 Section GI

Five pollen analyses were conducted on monolith/core GI1, mainly to date the flooding of the large central hummock. The peat above 32 cm was taken to be disturbed. The results (figure 46) correlate very well indeed with pollen diagram HI9, although the uppermost count at 14 cm may be correlated perfectly well at two different levels. The other four counts correlate more or less in sequence (33.5-63 cm GI1 equivalent to 97-132 cm HI9) and give the following dates and accumulation rates. The change from highly-humified, almost black peat, to pool-peat, occurred around 880 BC and the deep pool formed at this time filled in only slowly with a laminated greasy *Sphagnum cuspidatum* peat at a rate of 54.3 years/cm. This is the slowest accumulation rate found at Bolton Fell Moss and from it one may calculate that the pool existed for some 700 years or so (13.5 cm x 54.3 years/cm = 733 years). The calculation is not, of course, accurate, and depends upon mean accumulation rates extrapolated from the HI9 Age/Depth profile, but it does give some measure of the possible longevity of deep pools, a point noted elsewhere by Sjors (1946), and others working on surface ecology.

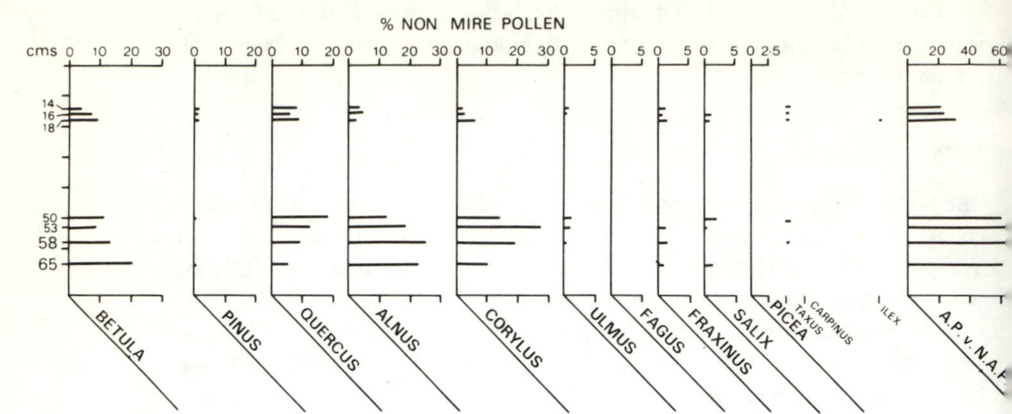

Figure 45. Pollen diagram from monolith DI1. Based on pollen sum of at least 250 non-mire pollen. Summary diagram showing main types only.

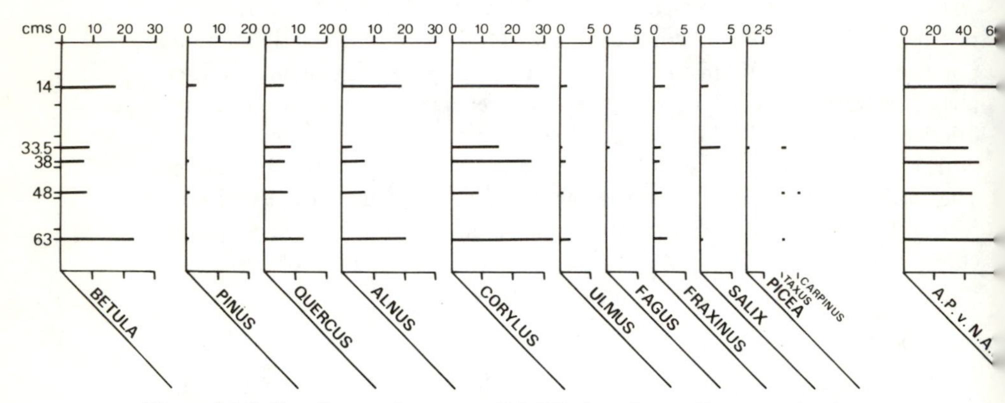

Figure 46. Pollen diagram from monolith GI1. Based on pollen sum of at least 250 non-mire pollen. Summary diagram showing main types only.

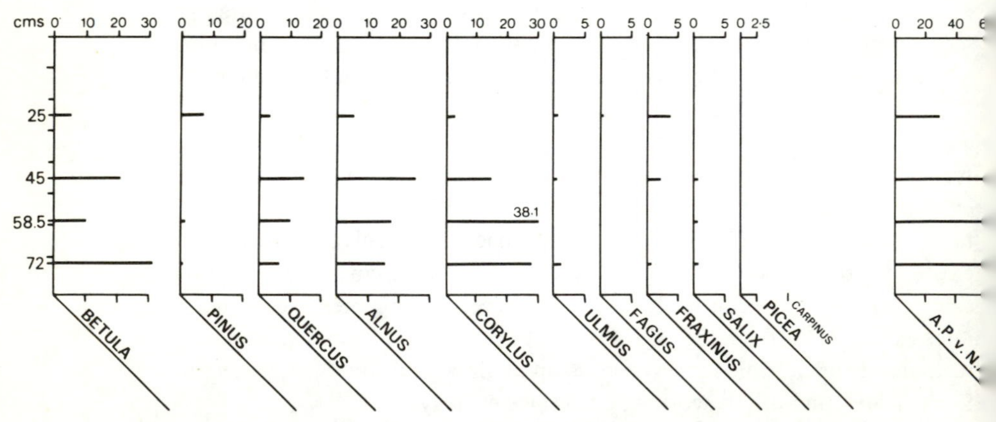

Figure 47. Pollen diagram from monolith GII1. Based on pollen sum of at least 250 non-mire pollen. Summary diagram showing main types only.

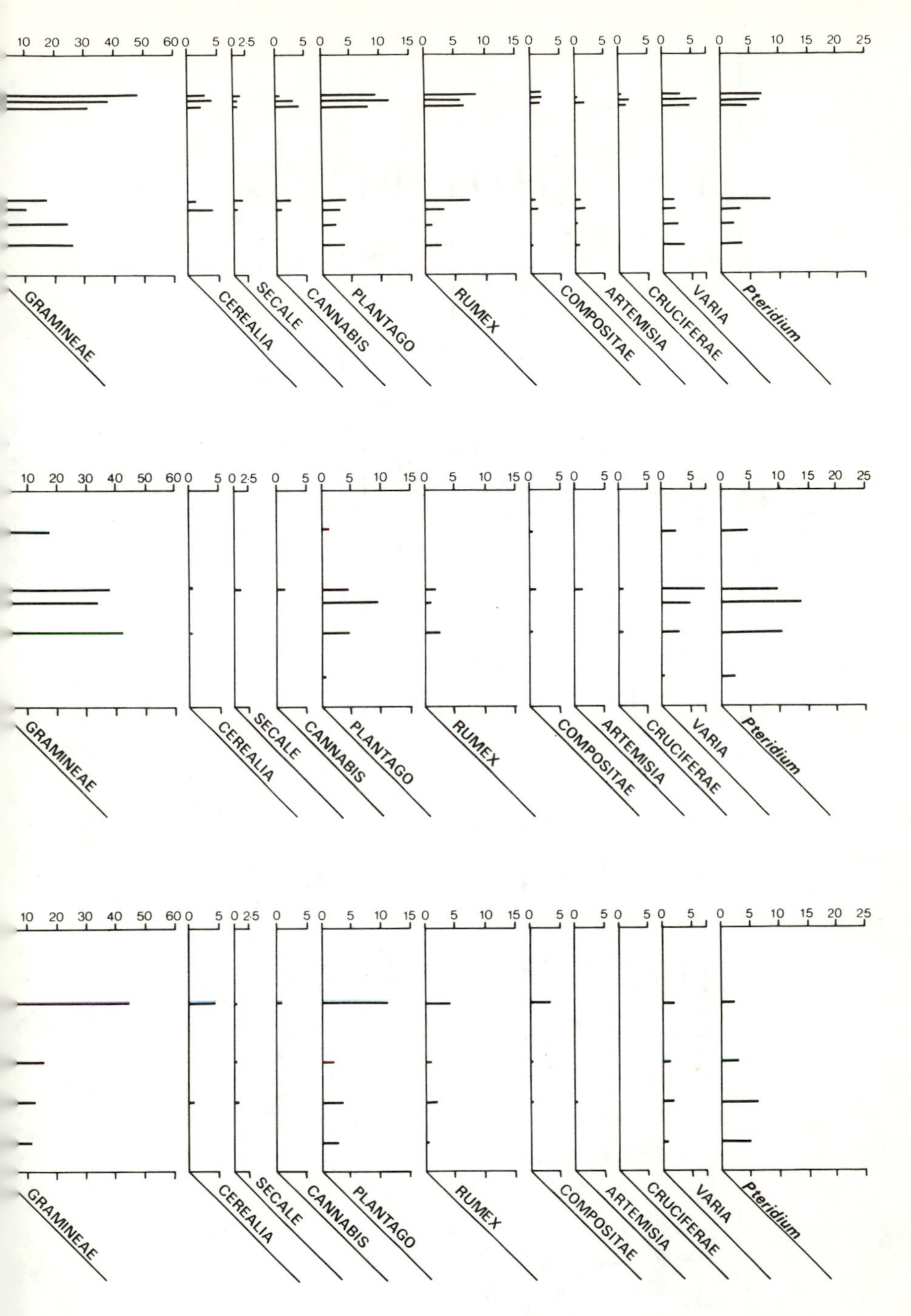

10 20 30 40 50 60
0 5
0 2·5
0 5
0 5 10 15
0 5 10 15
0 5
0 5
0 5
0 5
0 5 10 15 20 25
GRAMINEAE
CEREALIA
SECALE
CANNABIS
PLANTAGO
RUMEX
COMPOSITAE
ARTEMISIA
CRUCIFERAE
VARIA
Pteridium
10 20 30 40 50 60
0 5
0 2·5
0 5
0 5 10 15
0 5 10 15
0 5
0 5
0 5
0 5
0 5 10 15 20 25
GRAMINEAE
CEREALIA
SECALE
CANNABIS
PLANTAGO
RUMEX
COMPOSITAE
ARTEMISIA
CRUCIFERAE
VARIA
Pteridium
10 20 30 40 50 60
0 5
0 2·5
0 5
0 5 10 15
0 5 10 15
0 5
0 5
0 5
0 5
0 5 10 15 20 25
GRAMINEAE
CEREALIA
SECALE
CANNABIS
PLANTAGO
RUMEX
COMPOSITAE
ARTEMISIA
CRUCIFERAE
VARIA
Pteridium

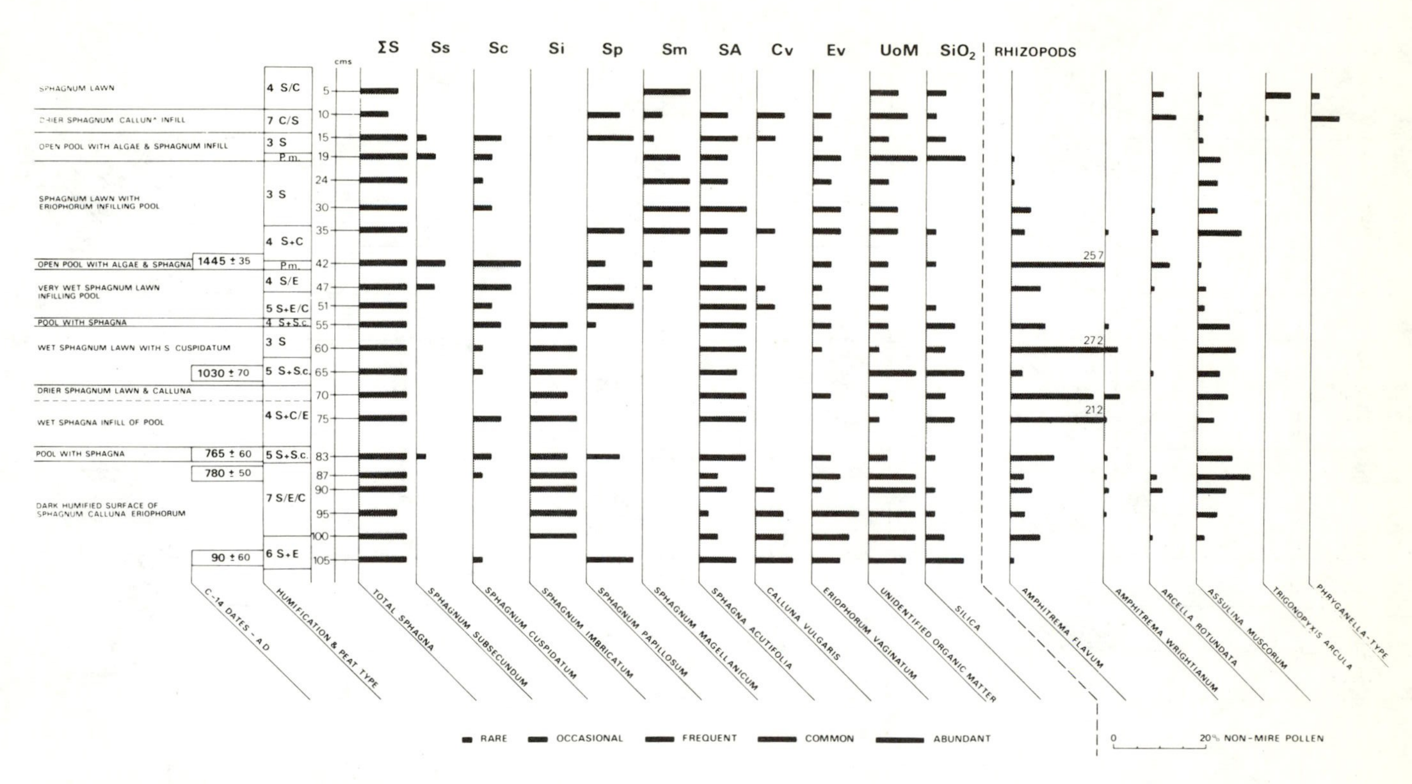

Figure 48. Macrofossil and rhizopod diagram from monolith HI9.

The moderately humified peat which grew all over the pool at about 50 cm depth is dated to around 150 BC and accumulated at a somewhat faster rate than the peat below; 20.3 years/cm between 48 cm (13 BC) and 38 cm (190 AD), slowing then to 35.5 years/cm up to 33.5 cm (350 AD).

The pollen count from 14 cm in GI1 is problematical. It was done before the others to test whether or not the upper peat was, as suspected, overturned. This suspicion was based on rather peculiar field stratigraphy, with merging, alternating bands of H3 and H5 peat (figure 18), the position of the site on the line of the old 1800 AD ditches, and on the detailed monolith stratigraphy which displays a banded mixture of black and brown peat, varying from H7 to H4, and varying from coarse to fine in texture. There was also a distinct break in the monolith peat, with the peat shrinking differentially either side of the break. When it was found that the pollen spectrum at 14 cm contained very high arboreal pollen and very little NAP (less than 20 % Gramineae), correlating best with HI9 at between 115 and 120 cm, it was concluded that the peat above the sharp boundary at 32 cm was overturned. While this remains a distinct possibility it has been found possible to correlate the 14 cm count with a level of 78 cm in HI9. The correlation is not as good as that at 117 cm HI9 but only inasmuch as the birch value was 10 % or so less than that in HI9. A third possibility exists in that the peat of 14 cm could come from a time period beyond the reach of 150 cm: HI9 (= 1440 BC), though with a similar pollen spectra. This is not as unlikely as it may at first appear since the pollen curves below about 110 cm do not vary much in overall composition and there are no reliable indicators such as the pine rise or hemp demise near the top of the diagram.

If the correlation at 78 cm is accepted the growth rate between 33.5 cm and 14 cm works out at 25.8 years/cm, with 14 cm dated around 855 AD and 0 cm in GI1 would then be around 1220 AD. The peat which accumulated after this date has then been removed and this is also plausible since, as explained, sections GI and GII are, in section, like two steps, GI being the lower.

Whichever correlation is accepted the stratigraphy above 32 cm: GI1 is by no means critical, especially since it only covers the period 400-1200 AD when few independent climatic correlations can be made, though it would mean that the change from H5 *Sphagnum-Calluna* peat to H3 *Sphagnum* peat at 18.5 cm would date to 740 AD – about the same time as the main humification change in HI9.

5.3.8 Section GII

The pollen diagram (figure 47) from this section comprises four spectra, spaced so as to date the main humification change and the pool-muds. The pollen correlation with HI9 was, in general, a very good fit on all the curves. The 72 cm

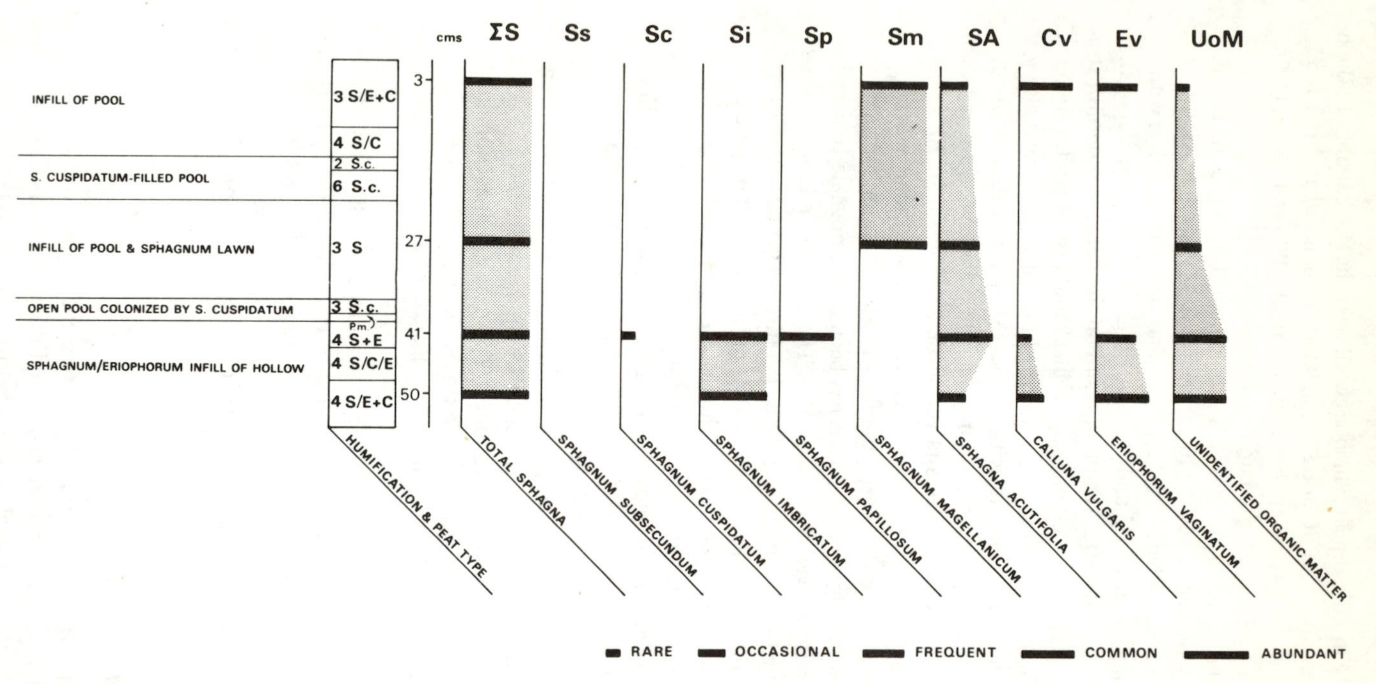

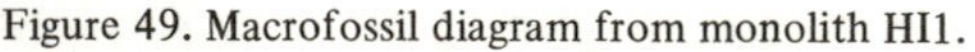
Figure 49. Macrofossil diagram from monolith HI1.

level in GII1 correlates at 77 cm: HI9 (875 AD) and so dates the main humification change to about 860 AD. Thereafter, in the hollow through which the monolith was taken (figure 19), peat accumulated rapidly (13.0 years/cm) until a pool formed at 60.5 cm (= 1025 AD). This pool lasted until 1050 AD (58.5 cm = 64 cm: HI9), and equates very well with the pool layer in HI9. Subsequent infill of the hollow, alongside a prominent hummock, proceeded at a rate of 21.5 years/cm (45 cm = 47 cm: HI9 = 1340 AD) until another marked pool formation at 38.5-36 cm (1485-1540 AD). This pool level, only slightly later than that in HI9, flooded the hummock formation which had until then dominated the stratigraphy of section GII, and a widespread lawn of H3 *Sphagnum* then accumulated at 22.3 years/cm until the final pool formation at 27 cm (1745 AD), into which was tipped the spoil from the 1800 AD ditch.

5.4 MACROFOSSIL ANALYSES

The results described here have been achieved using the standardised sieving method described earlier. Some analyses were performed at an early stage of the research on monoliths from sections A and B but the standardised method, used from 1972 on, superseded these results and they are not reported here; they do not in any case differ substantially from the later results. As with the pollen analyses these results are reported not in chronological order but taking the closely-sampled HI9 first, then the other monoliths of HI, followed by the results from sections A-G.

5.4.1 Section HI, monolith 9

The lowest five samples on this diagram (figure 48) are in accord with the field stratigraphic evidence of a well-humified peat, which accumulated relatively slowly between 90 and 780 AD. Although *Sphagnum* remains are common or abundant they are not very well preserved and the proportion of 'unidentified organic matter' is very high, along with high levels of *Calluna* and *Eriophorum. Sphagnum imbricatum* dominates the moss flora, though it is absent from the lowest sample at 105 cm, where its place is taken by *Sphagnum papillosum.* The main humification change between the samples at 87 and 83 cm is well brought out in the macrofossil results. The *Sphagnum* flora becomes a mixture of species, *S. subsecundum, cuspidatum, imbricatum* and *papillosum,* together with probably more than one species in the Sphagna Acutifolia aggregate group, growing together in a pool environment to judge from the greasy character of the deposit as well as the remains of the first two species above. The hygrophilous rhizopods, mainly represented by *Amphitrema flavum* and *Assulina muscorum,* also show increased frequencies at this level, while both *Calluna* and *Eriophorum* decline dramatically – the former species does not reappear for some 30 cm or so.

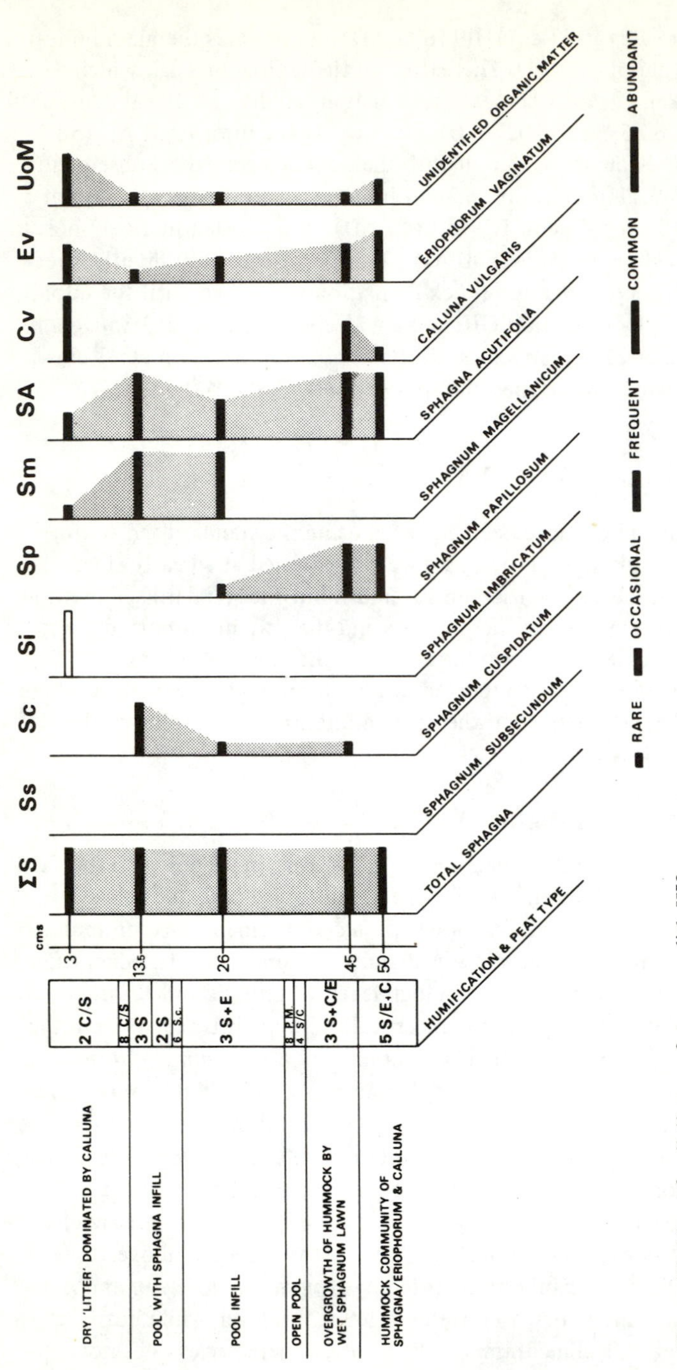

Figure 50. Macrofossil diagram from monolith HI2.

The wet lawn conditions which obtained after about 770 AD are interesting in that *S. imbricatum* maintains more or less its former abundance, in competition with *S. cuspidatum* and much acute-leaved Sphagna. This argues for a wide tolerance range with regard to water levels or for a rapid change to a more aquatic 'ecad' of the species. Such flexibility was, however, apparently to no avail in what is perhaps the major event to be observed in this and many of the other macrofossil diagrams, the extinction of *S. imbricatum* and the taking over of its role as the main peat former by *S. papillosum* and *S. magellanicum.*

The change is often accompanied by a marked stratigraphic change (see results from HI4 for example) but in the hollow situation of HI9 it simply coincides with a minor pool layer at 55 cm. In this pool *S. imbricatum* still makes up some 50 % or so of the *Sphagnum* remains, with *S. cuspidatum* frequent and a few leaves and branches of *S. papillosum,* as well as the ubiquitous Sphagna Acutifolia. A few centimetres higher at 51 cm (dated to the early 1200's) *S. imbricatum* has vanished completely, even though the other indicators point to slightly drier conditions. Its role is taken over by *S. papillosum,* joined a little later by the first appearance of *S. magellanicum.* These species are held in check by the major pool development of the first half of the 15th century which laid down a thick yellow-green pool-mud of algal remains and abundant *S. cuspidatum,* but thereafter *S. magellanicum* becomes more and more important until from 35 cm onwards it pushes *S. papillosum* into a secondary role from which it only emerges sporadically, possibly in response to drier conditions as between 15 and 10 cm in diagram HI9. From the evidence of the pools at 55 cm and 19 cm it would seem that the effect of flooding was to submerge the existing tops of the cymbifolian Sphagna, *S. imbricatum* and *S. magellanicum* being abundant elements of these respective pools, and then the pool infill phase sees these species unable to compete with, for example, *S. cuspidatum.* The infill of the 42 cm pool would appear to be a special case, the hygrophilous species of Sphagna and of rhizopods being largely absent and suggesting a rather drier lawn. The place of *S. papillosum* in this pool infill succession is an interesting one and accords with the author's observations of its behaviour on a number of British bogs, particularly on Tregaron Bog, and with Ratcliffe & Walker's (1958) observations on the Silver Flowe bogs.

The topmost samples from monolith HI9 also show an interesting change in the rhizopod fauna with the replacement of the *Amphitrema-Assulina* association by a 'tyrfoxene' association of *Arcella-Trygonopyxis-Phryganella* species typical of drained or 'dying' raised bogs according to Tolonen (1971). Finally it is interesting to note that the frequency curve of non-opaline silica shows a close coincidence with clearance episodes in the pollen diagram HI9, with peaks of wind-blown SiO_2 corresponding to the Roman clearance (105 cm), the Norse clearance (65 cm) and the Napoleonic clearance (19 cm).

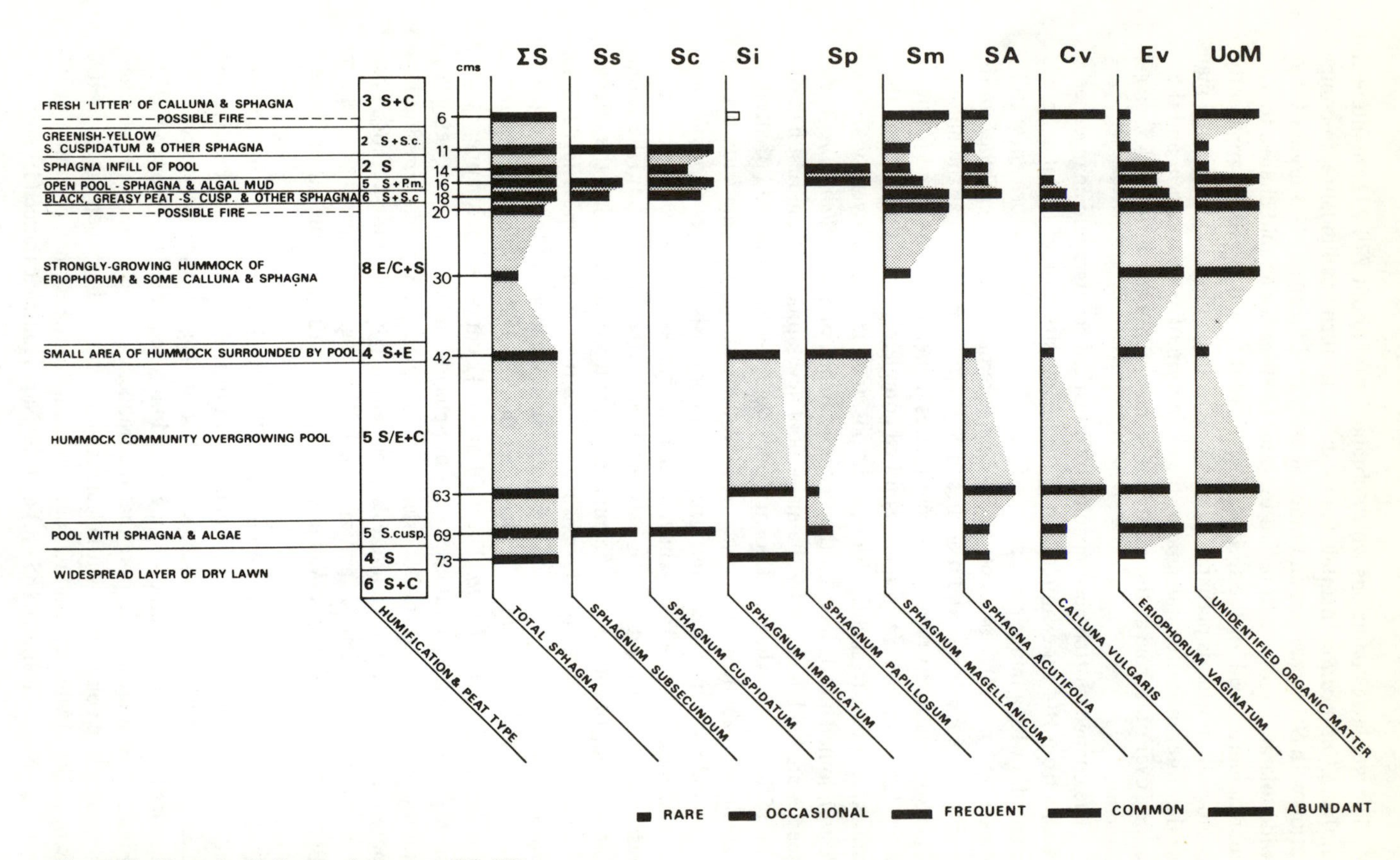

Figure 51. Macrofossil diagram from monolith HI3.

The overall effect of this macrofossil diagram is to emphasise a succession of Sphagna resulting in a change in the dominant species, and to characterise the pool and wet lawn phases as being associations of species rather than monospecific replacements of the previous vegetation. Also, considering the radiocarbon and pollen dating of monolith 9, the time-scale of species interplay here demonstrated is markedly out of step with any cyclic theory that can be of relevance in post-glacial terms.

5.4.2 Section HI, monoliths 1 and 2

These two monoliths may be taken together because of their similar stratigraphic position (figure 20), both covering the middle and upper pool layers. The macrofossil results (figure 49 and 50) show that in monolith 1 the hollow to the left of the persistent hummock filled in with a *Sphagnum imbricatum-papillosum*-Acutifolia mixture, in which *Calluna* and *Eriophorum,* immigrating from the surrounding hummocks, were important constituents. The correspondence with the field stratigraphy is high. After the pool phase, the samples at 27 and 3 cm show the dominance of *S. magellanicum* and the disappearance of *S. imbricatum* and *papillosum.*

A similar story is evident from monolith 2, except that by the time of the first sample at 50 cm, *S. imbricatum* had already died out – the sample at 45 cm, just above the humification change, is dated to 1290 AD by pollen correlation. The *S. papillosum*-Acutifolia association of the lower part of the diagram is followed by *S. cuspidatum* and then *S. papillosum* is again ousted by *S. magellanicum* in post-Medieval times. The upper pool of circa 1800 AD does nothing to diminish the abundance of *S. magellanicum,* the uppermost count at 3 cm being contaminated by humified *S. imbricatum* from the peat stacks left to dry on the surface.

Together these two monoliths confirm the broad trends visible in monolith HI9 and there is nothing in these results to support cyclic changes over at least the last 700 years.

5.4.3 Section HI, monoliths 3, 3a, 4, 5, 5a, 6 and 7

The macrofossil results from these monoliths hang together as a suite taken from the central pool and hummock complex of section HI. Three of the monoliths have been analysed in some detail (3, 4 and 6) as they come from the crucial areas of apparent cyclic succession within the complex; the other monoliths were analysed to give supporting detail.

Monolith 3 has a most interesting succession (figure 51), beginning with a rapid change from humified surface to pool, and then to hummock between 73 and 63 cm. The growth rate at 13.75 years/cm is identical to that of HI9 over the same period and means that the pool layer from 71-67 cm lasted 55 years. In that short time the surface changed from one dominated by *S. imbri-*

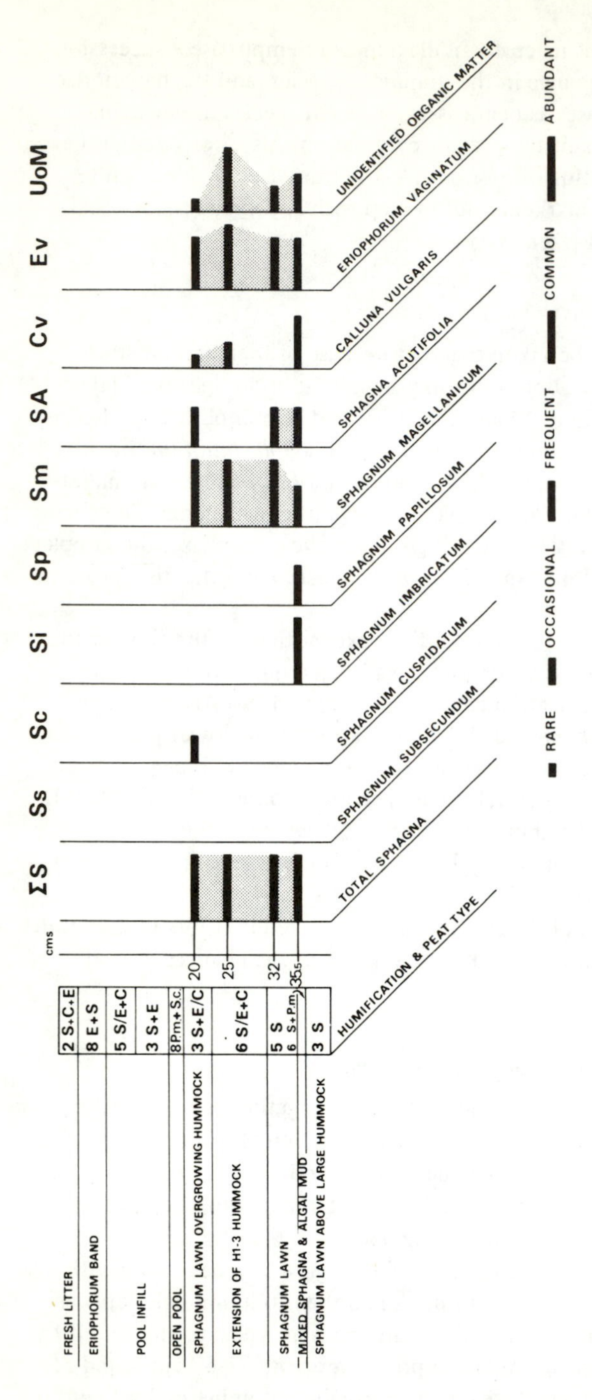

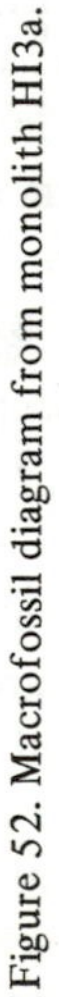

Figure 52. Macrofossil diagram from monolith HI3a.

catum to one in which that species was not represented at all (*S. subsecundum* and *S. cuspidatum* dominance of the 69 cm sample) and back again. The hummock was then dominated by a mixture of *S. imbricatum* and *S. papillosum* (the latter species having characteristically entered at the pool stage) for some 350 years, up to the 42 cm level, dated around 1380 AD. At this point the hummock was almost overwhelmed by the middle pool; despite the absence of *S. cuspidatum* on the macrofossil diagram this can be picked up in the lowered values of the four right hand curves (S. Acutifolia, *Calluna, Eriophorum* and unidentified organic matter) and in the heightened value of *S. papillosum* compared with *S. imbricatum.* However, the knot of hummock species remained and grew up to form a vigorous hummock (growing at more or less the same rate as the hollow infill in HI9, circa 15 years/cm) composed primarily of *Eriophorum vaginatum* with occasional remains of *S. magellanicum* (sample at 30 cm, figure 51).

This hummock also expanded to the right of the stratigraphic section (figure 20) and monolith 3a was taken to investigate this. The macrofossil diagram (figure 52) shows, not unexpectedly, a lower level containing *S. imbricatum* as well as *S. papillosum* and *S. magellanicum* which is then eclipsed by the *S. magellanicum/Eriophorum vaginatum* association of the spreading hummock between about 1500 and 1700 AD.

The overgrowth of this hummock was analysed in detail by the upper macrofossil samples of HI3. At 20 cm the hummock top, a well-humified mixture of *S. magellanicum, E. vaginatum* and *Calluna vulgaris,* appears perfectly capable of further growth – there is no S. Acutifolia present, such as *S. rubellum,* typical of the highest levels of hummock development, and the *Calluna* twigs were small (about 2 mm average diameter), suggesting that the hummock was not degenerating. (The 'possible fire' legend at this level refers to small pieces of charcoal found during the analyses). The next sample, at 18 cm, similarly shows abundant *S. magellanicum* but also much *S. cuspidatum* and *subsecundum* with some Sphagna Acutifolia thought to be mainly *S. tenellum.* The boundary at 19 cm is dated at 1740 AD by correlation with HI9. By 16 cm the *S. magellanicum* values have fallen and increased values of *S. subsecundum* and *cuspidatum,* accompanied by abundant *S. papillosum,* are mixed with algal mud in a pool dated pollen-analytically to 1785 AD. At 14 cm *S. subsecundum* is no longer present, though the *S. cuspidatum* content of the fresh peat attests to the moist conditions of formation, which become wetter again at 11 cm, a sample which once more contains abundant *S. subsecondum* and dated to 1825 AD on correlation with the HI9 Age/Depth profile, before the drier conditions of the recent *S. magellanicum/Calluna*-dominated surface. The overgrowth of this marked hummock is then to be seen not in terms of a degenerating, dry hummock-top, devoid of Sphagna, but in the context of a closely-dated hydro-climatological event which affected hummocks and hollows equally.

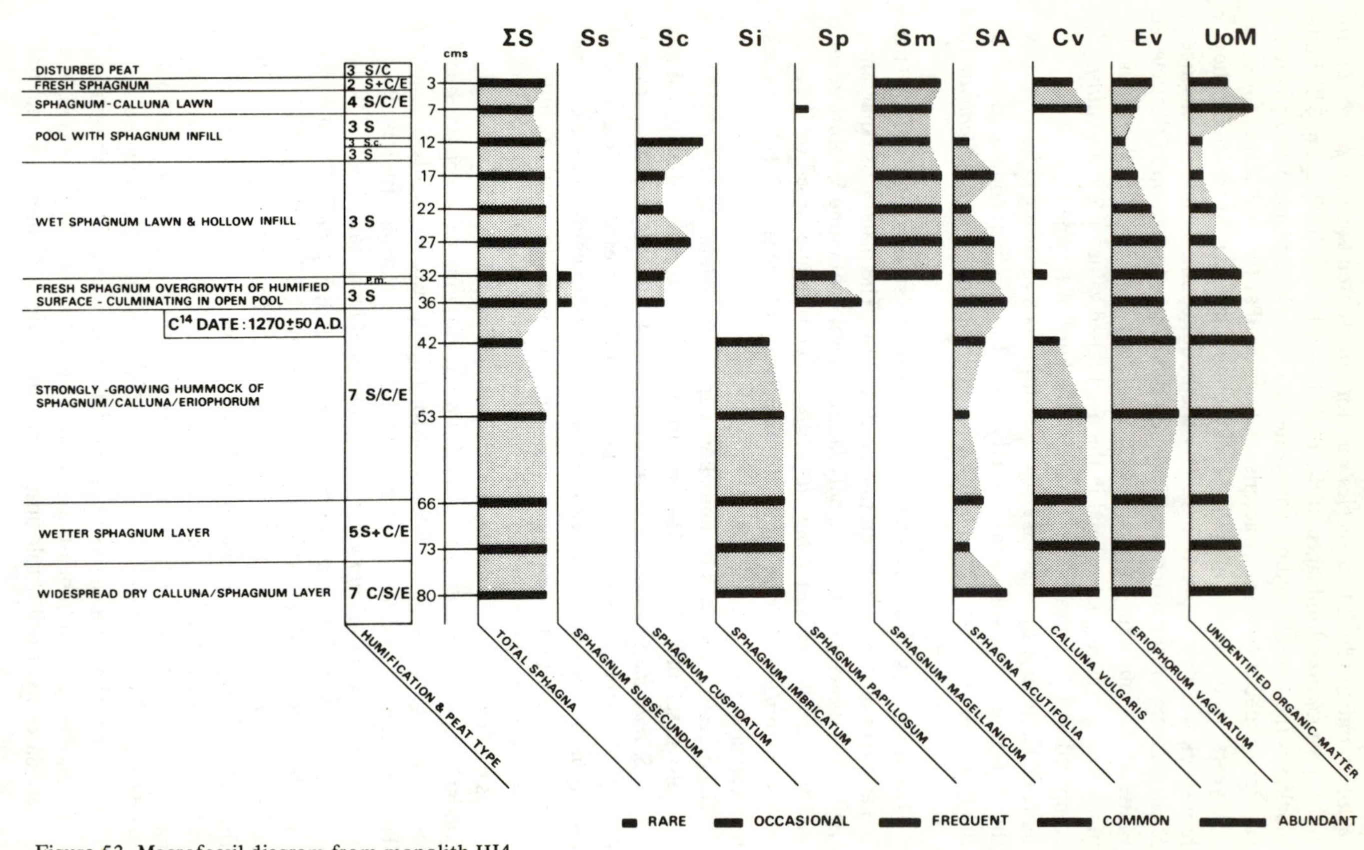

Figure 53. Macrofossil diagram from monolith HI4.

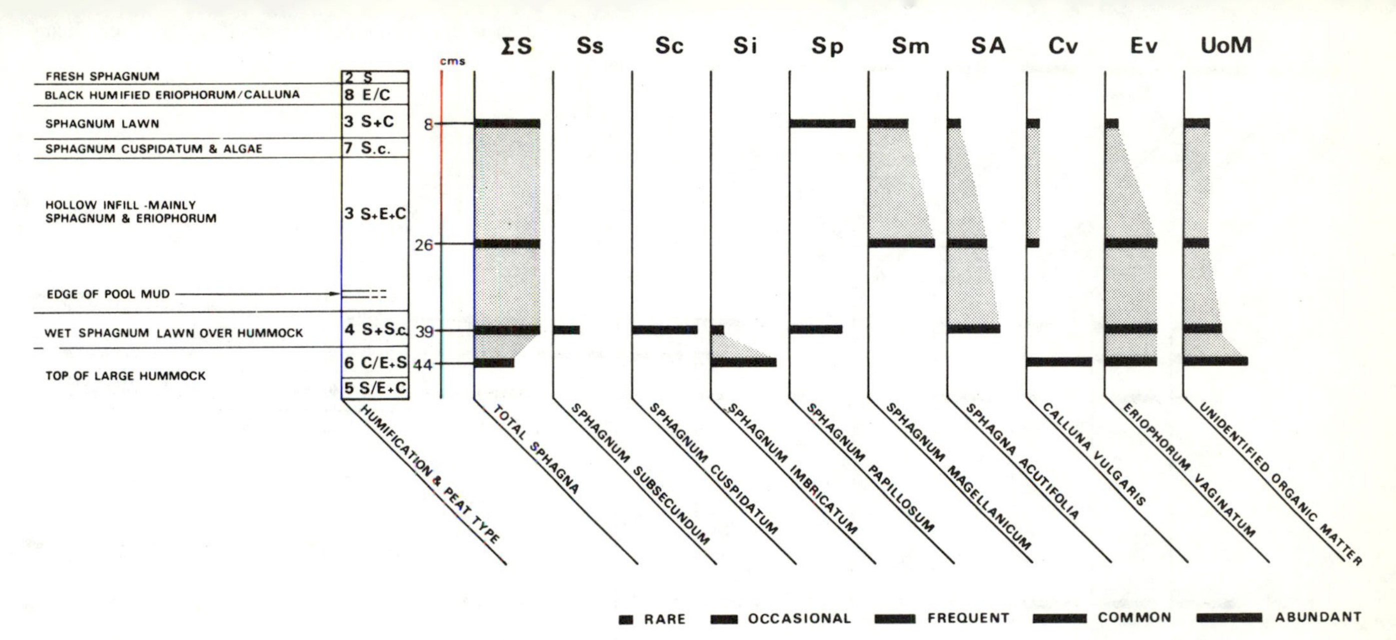

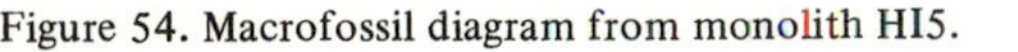
Figure 54. Macrofossil diagram from monolith HI5.

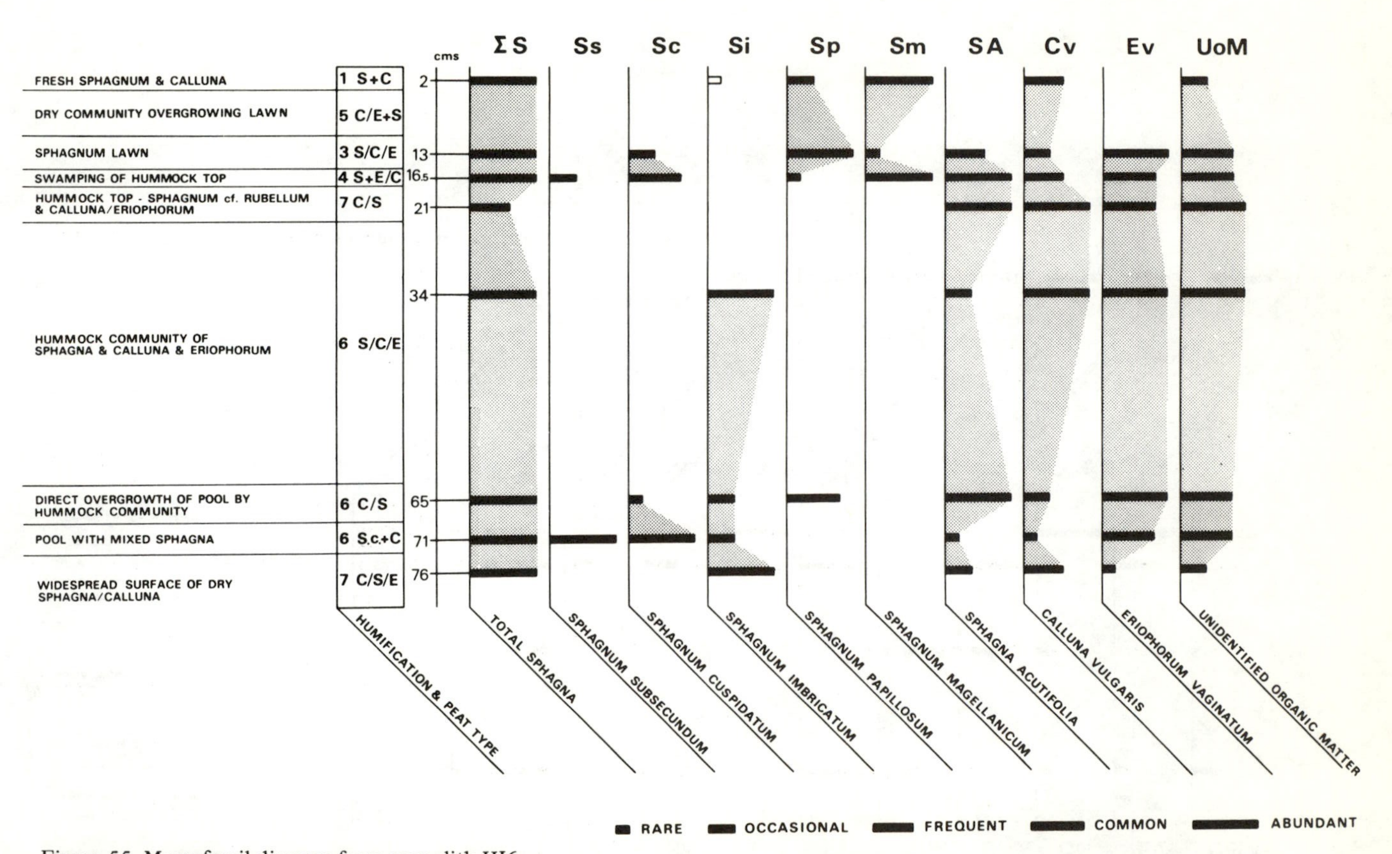

Figure 55. Macrofossil diagram from monolith HI6.

The macrofossils from monolith 4 (figure 53) are as straightforward as the two-fold stratigraphy of the monolith. The basal hummock section, undeflected by the minor wetter layer between 66-75 cm, is very much dominated by *S. imbricatum,* with abundant *E. vaginatum* and *Calluna.* Amounts of Sphagna overall fall towards the top of the humified layer and Sphagna Acutifolia gain at the expense of *S. imbricatum.* The upper unhumified section displays a modified version of the HI9 model with the *S. papillosum/magellanicum* sequence taking over from *S. imbricatum,* with *S. cuspidatum* indicating the wet nature of the *Sphagnum* lawn. The upper pool at 12 cm depresses the *S. magellanicum* values only slightly, and *S. papillosum* once more makes an appearance at the infill stage. The rate of infill of this hollow, from the pool-mud at 32 cm to the surface, works out at 15.9 years/cm, more or less identical to that of the HI3 hummock.

The monolith 5 results (figure 54) confirm the above succession, even down to the appearances of *S. papillosum* at more or less the same depths.

Monolith 6, and the associated monoliths 5a and 7, give a comprehensive view of the development of the eastern end of this central hummock-pool complex. The lower pool sequence is remarkably similar to that in monolith 3 (compare figures 51 and 55) with once more the direct overgrowth of the pool, similarly rich in *S. subsecundum* and *cuspidatum* but in which some *S. imbricatum* survives, by a hummock community with a *S. papillosum* precursor. The composition of the HI6 hummock is well shown in the 34 cm sample, with abundant *S. imbricatum, Calluna* and *E. vaginatum.* The hummock top sample (21 cm) is interesting in that *S. imbricatum* has now died out (some time between 34 and 21 cm; 1425-1660 AD) to be replaced by Sphagna Acutifolia, probably *S. rubellum,* and *Calluna.* This 'mature' hummock top was swamped at about 1740 AD by a mixed, wet *Sphagnum* lawn of at least five species including the most hygrophilous but also *S. magellanicum* (see figure 55). Once more the *S. papillosum* to *S. magellanicum* sequence unfolds up to the present surface.

The supplementary information from monolith 7 (figure 56) is valuable in confirming that the lower hummock contains only *S. imbricatum* and Sphagna Acutifolia, being dominated by *Eriophorum vaginatum,* and in showing that the overgrowth of this hummock did not involve *S. cuspidatum* but merely a change to a purer, less humified *S. imbricatum* lawn. This underlines the 'ecological plasticity' of this important species, a fact noticed by others, particularly Green (1968).

Monolith 5a, from the left of the HI6 hummock, is of interest in showing, for the first time, a mixture of *S. imbricatum, papillosum* and *magellanicum,* with *papillosum* dominant (figure 57). This sample from 27 cm is of the outgrowth of the main hummock (figure 20), and dated to 1670 AD. It demonstrates that *S. imbricatum,* which died out between 1270 and 1340 AD in monolith HI4 was able to 'hang on' on the HI6 hummock, and then capable

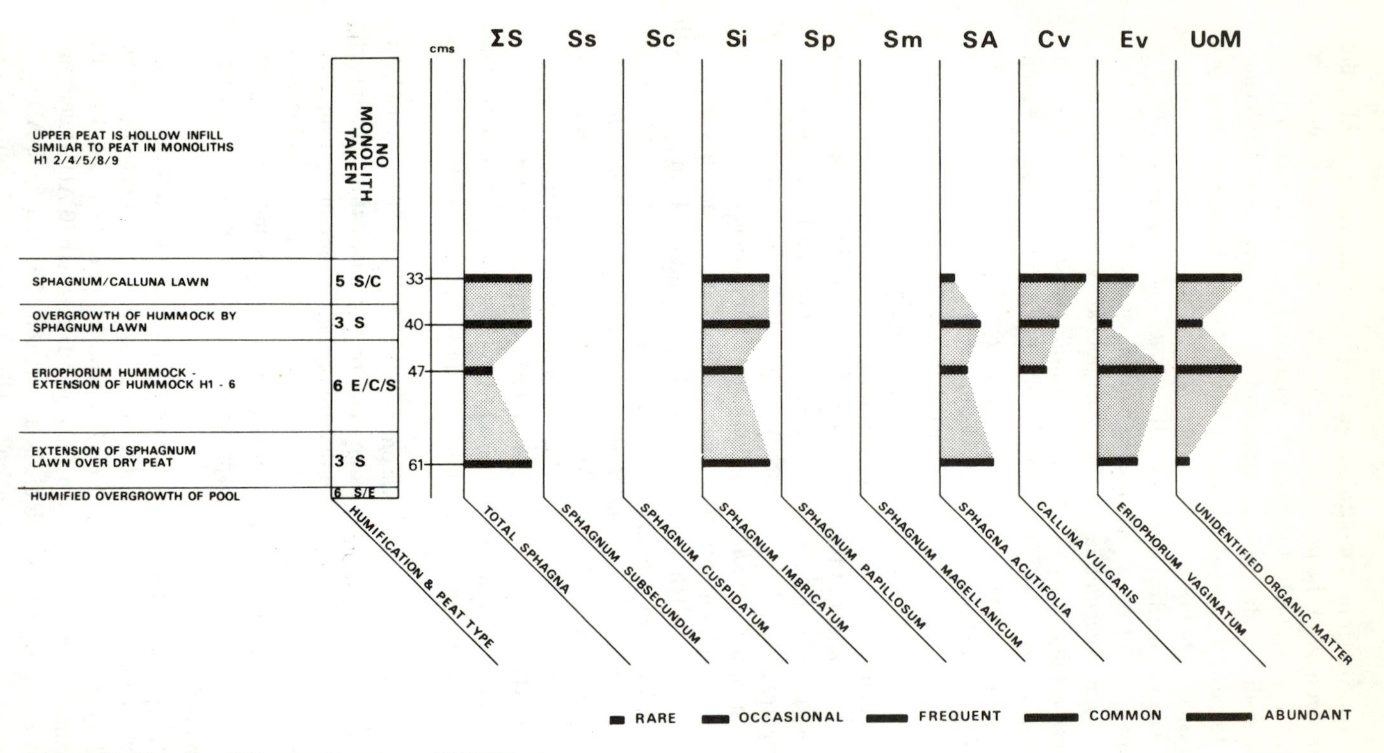

Figure 56. Macrofossil diagram from monolith HI7.

of a limited recolonisation of the hollow between HI3 and HI6 some 330 years later! Thereafter the succession in monolith 5a follows the established pattern to an almost pure *S. magellanicum/Calluna* community.

With these seven macrofossil diagrams, dated by the pollen-correlative technique, we can now see that a succession which on the basis of the field stratigraphy between 200 and 420 cm would be held to support the theory of cyclic alternation of hummocks and pools, shows instead a remarkable degree of simultaneous species change and succession which, as will be shown in the next section (5.5) can be satisfactorily related to independently-known climatic changes.

5.4.4 Section HI, monolith 8

The results from this monolith (figure 58) show a sudden flooding of a *Sphagnum imbricatum-Eriophorum* surface at 965 AD, followed by centuries of growth of a very wet nature – *S. subsecundum* and *cuspidatum* persisting through the infill stage, *Calluna* and *Eriophorum* remains very sparse – during which period *S. imbricatum* is replaced by *S. papillosum.* An interesting point to note in this case is that the extinction of *S. imbricatum* does not appear to be associated with any dramatic stratigraphic change such as occurs in monolith HI4. The species appears to come back strongly after the pool phase, though this could be due to re-immigration from the hummock extension shown on figure 20, but then dies out somewhere between 60 and 45 cm, corresponding to 1160 AD and 1380 AD. While its demise may simply be a delayed reaction to the pool phase, or else due to some climatic effect which did not quite register on the stratigraphy of monolith 8, the simpler explanation in this case is that it was supplanted by successful competition from *S. papillosum* – perhaps aided by the wet conditions.

Sphagnum papillosum and Sphagna Acutifolia continued to build up the profile, which passed through a pool stage exactly in sequence with that of HI9, only 1.5 m away, and then *S. papillosum* built up the small 'boss' of lightly humified, almost pure moss peat over the pool. This interesting feature, 4 cm high, took approximately 40 years to build and appears then to have merged with a rapidly accumulating mat of *S. magellanicum* which infilled any remaining hollow, and was then joined by *Calluna* and *Eriophorum* in a rapid build-up of a hummock extending to the surface. Throughout this monolith it is apparent that we are dealing with a succession, probably dependent on climate acting through water level, rather than a cycle.

5.4.5 Sections AI and AII

The three monoliths, AI1, AI2 and AII1, produced two macrofossil profiles which agreed with the general trends of the section HI results, and one which

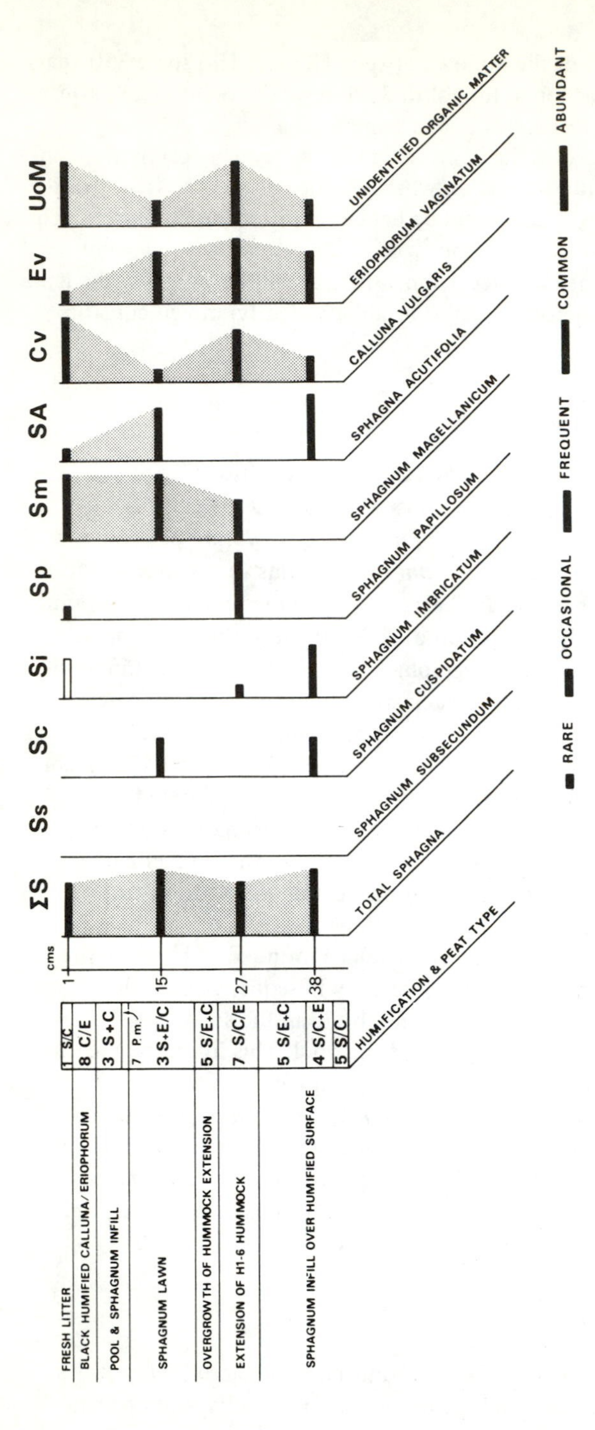

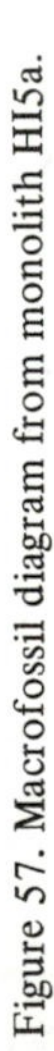

Figure 57. Macrofossil diagram from monolith HI5a.

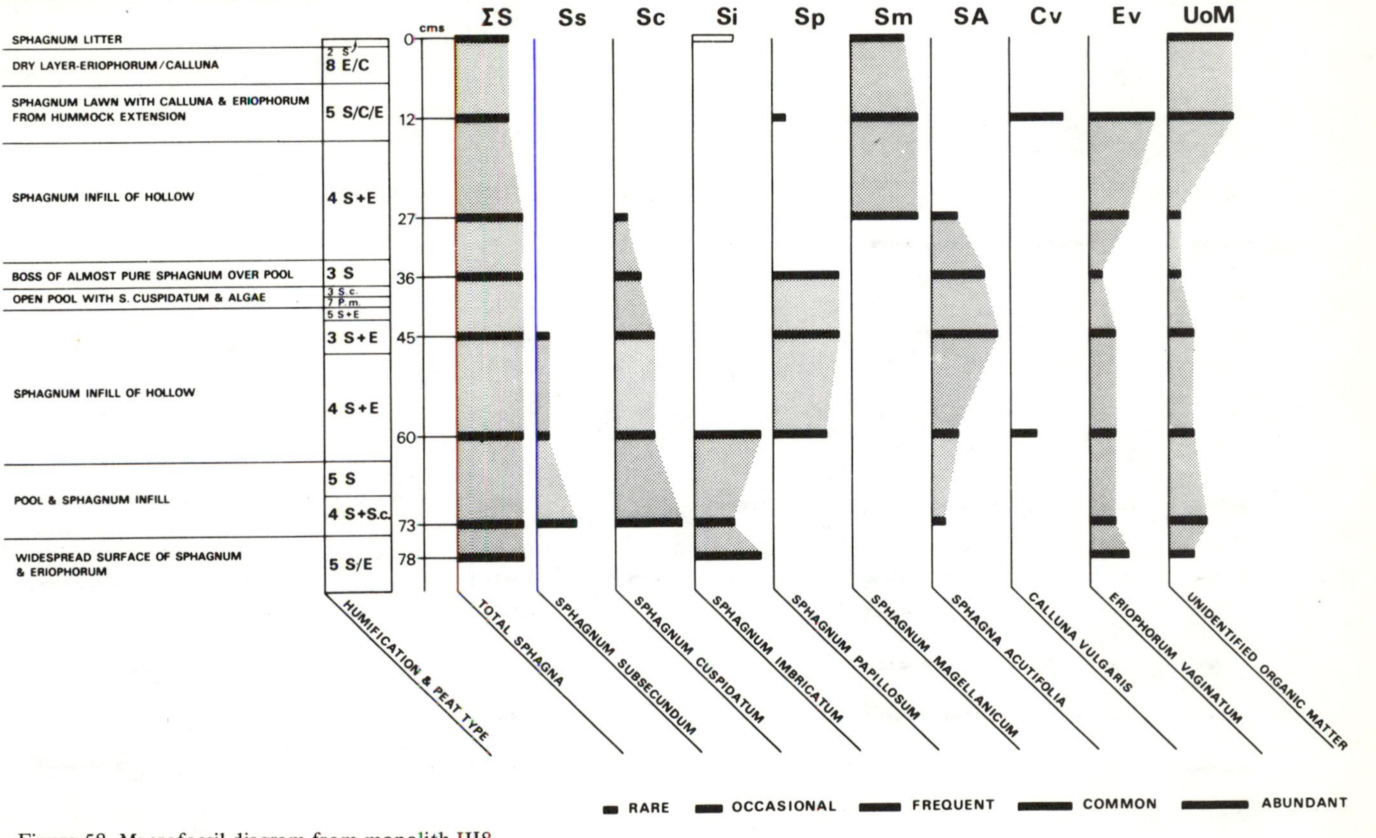

Figure 58. Macrofossil diagram from monolith HI8.

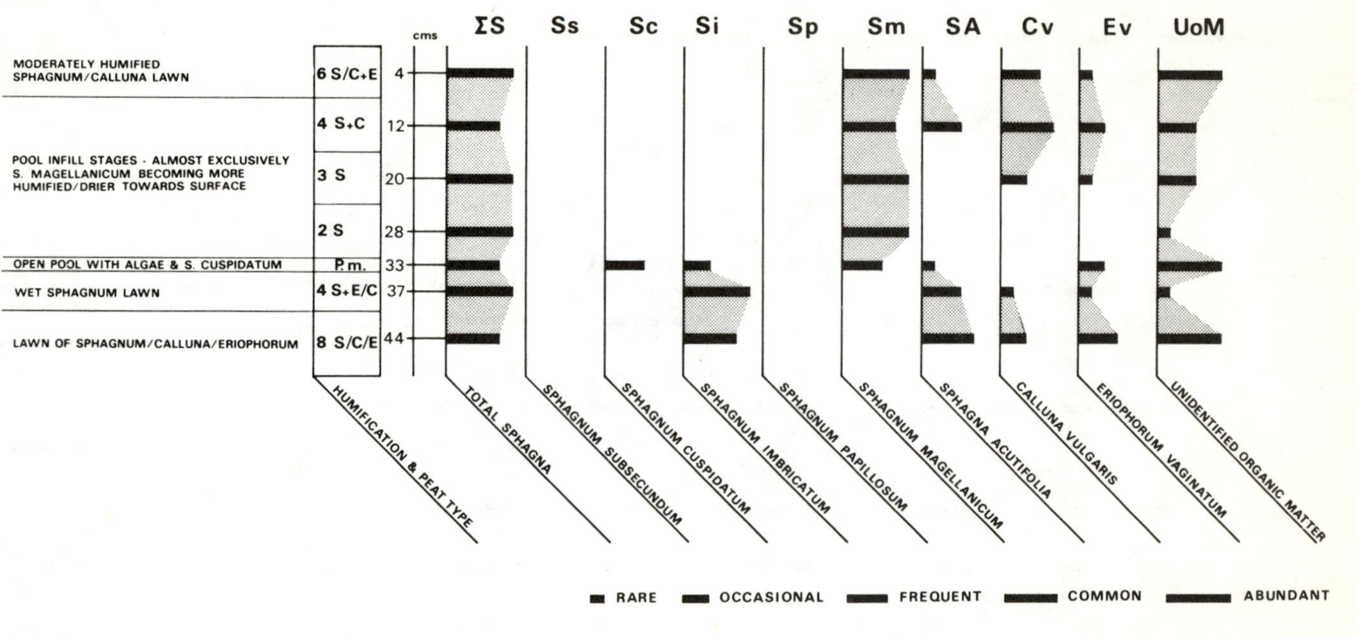

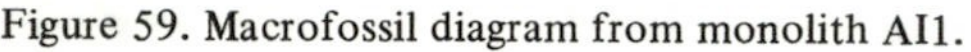
Figure 59. Macrofossil diagram from monolith AI1.

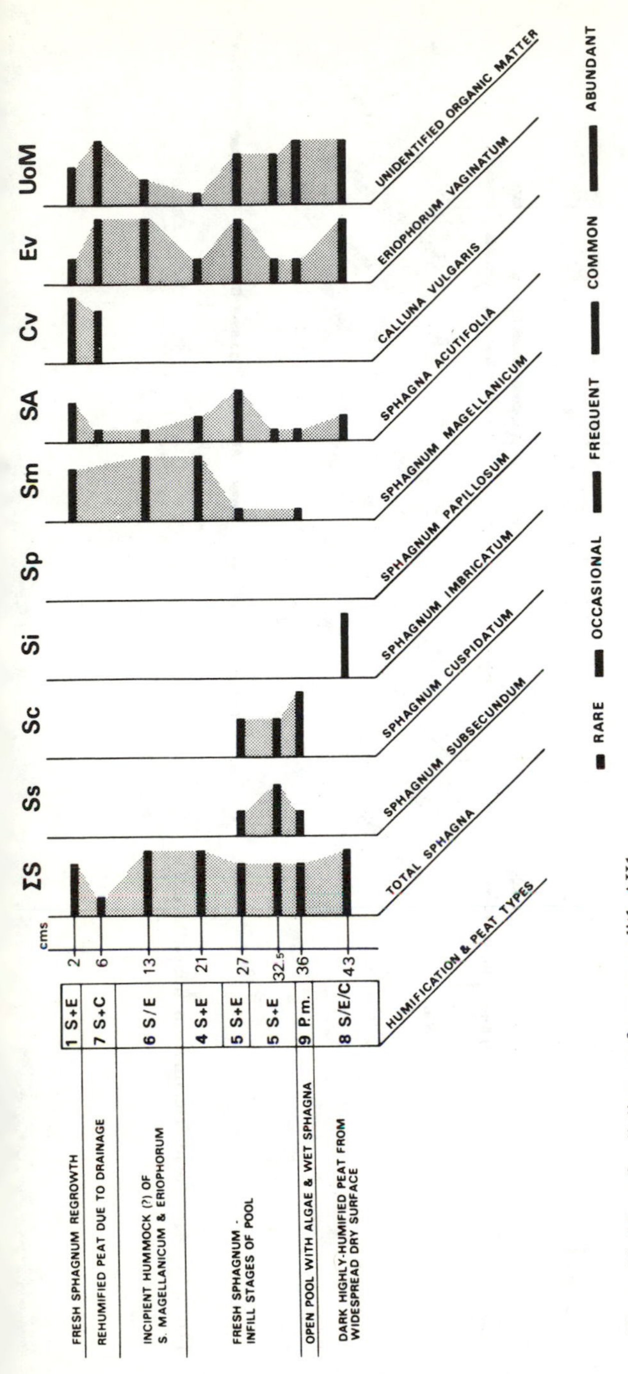

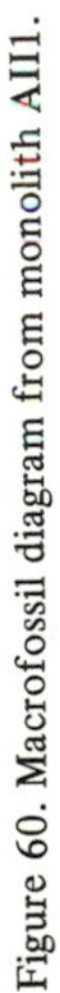

Figure 60. Macrofossil diagram from monolith AII1.

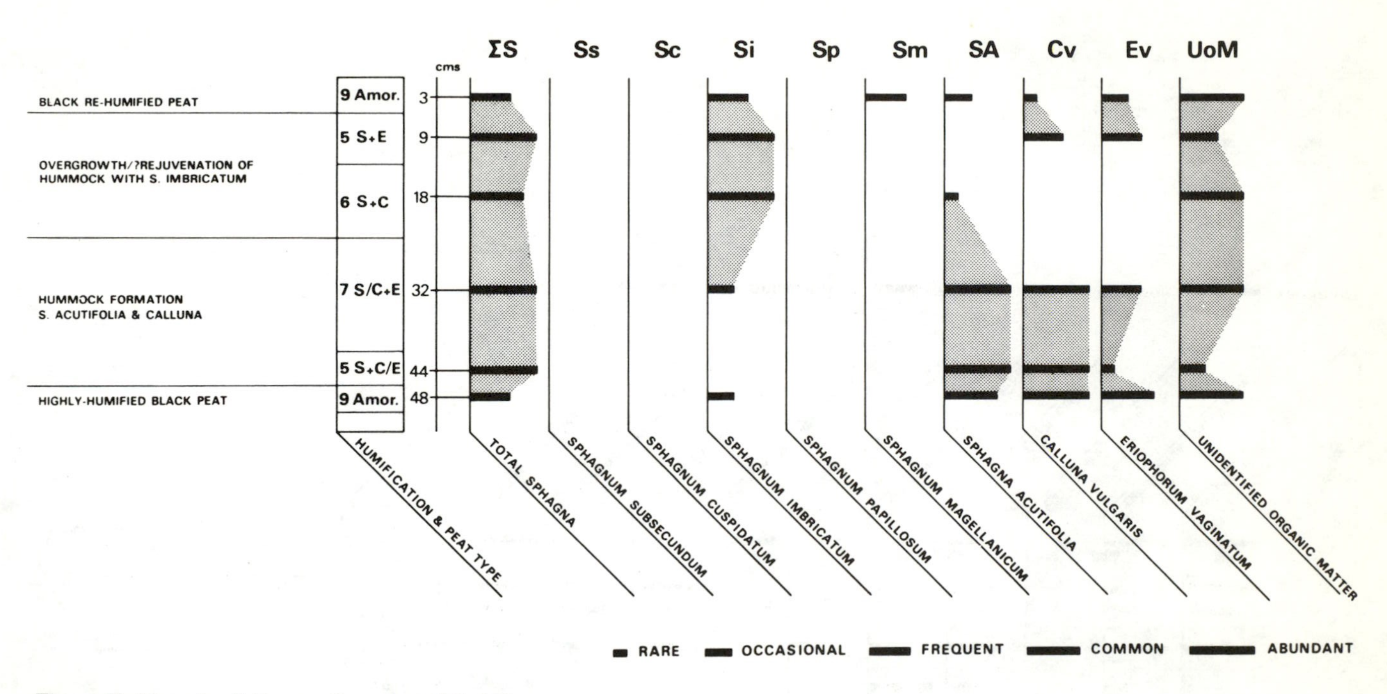

Figure 61. Macrofossil diagram from monolith AI2.

went against this trend. The diagrams from AI1 (figure 59) and AII1 (figure 60), from pool profiles, agree in showing a basal surface of *S. imbricatum*-dominated vegetation which is extinguished by the pool phases of the 1315-1420 AD period. Then, without the intervention of any *S. papillosum,* the pool infill quickly became dominated by *S. magellanicum* with varying amounts of *Eriophorum* and other species. One particular count, 28 cm in AI1, was exclusively *S. magellanicum* of low humification, whereas the infill of the AII1 pool involved a mixture of *S. subsecundum, cuspidatum* and other species before *S. magellanicum* could achieve dominance.

The results from monolith AI2 (figure 61) are very different and show that *Sphagnum imbricatum* survived up to at least 1800 AD on Bolton Fell Moss. It is clear that although it was present in the hummock, the formation was dominated by Sphagna Acutifolia (cf. *rubellum*) *Calluna* and some *Eriophorum vaginatum.* However, the rejuvenation of the hummock and its incorporation into a moderately humified surface from 24 cm upward, dated to 1420 AD, led to almost total domination by *S. imbricatum* – see for example the 18 cm count in figure 61 – a phenomenon very much in tune with Osvald's observations (see figure 4). At the 3 cm level we see the first showing of *S. magellanicum* and one may speculate on whether or not, had the section grown up to the 1950's as did section B1, described next, this species would have ousted *S. imbricatum.* In any case it can be seen from these results that since Medieval times at least *S. imbricatum* has persisted in hummock locations after being wiped out in lawn and hollow situations.

5.4.6 Section BI

The only monolith from this section yielded a not unexpected result (figure 62). The hummock which stretches from 50-27 cm is primarily of *S. imbricatum, Calluna* and *Eriophorum,* whereas the overgrowing lawn of around 1670 AD is of *S. magellanicum.* This species manages to withstand the very wet phase (with *S. subsecundum* as well as *S. cuspidatum*) around 1800 AD to form a dry lawn community with *Calluna* and other Sphagna during the 19th and early 20th centuries.

It is noticeable that *S. papillosum* is absent from all four profiles from areas A and B, but apart from its chance absence from the south-east quadrant it is difficult to advance a plausible reason for this, especially in the general absence of autecological data on the species, a difficulty which applies to most species of *Sphagnum,* as various authors have noted (Dickson 1973).

5.4.7 Section CI

The three monoliths from this section all show somewhat mixed, discontinuous macrofossil profiles. Monolith CI1 (figure 63) from the left of the section,

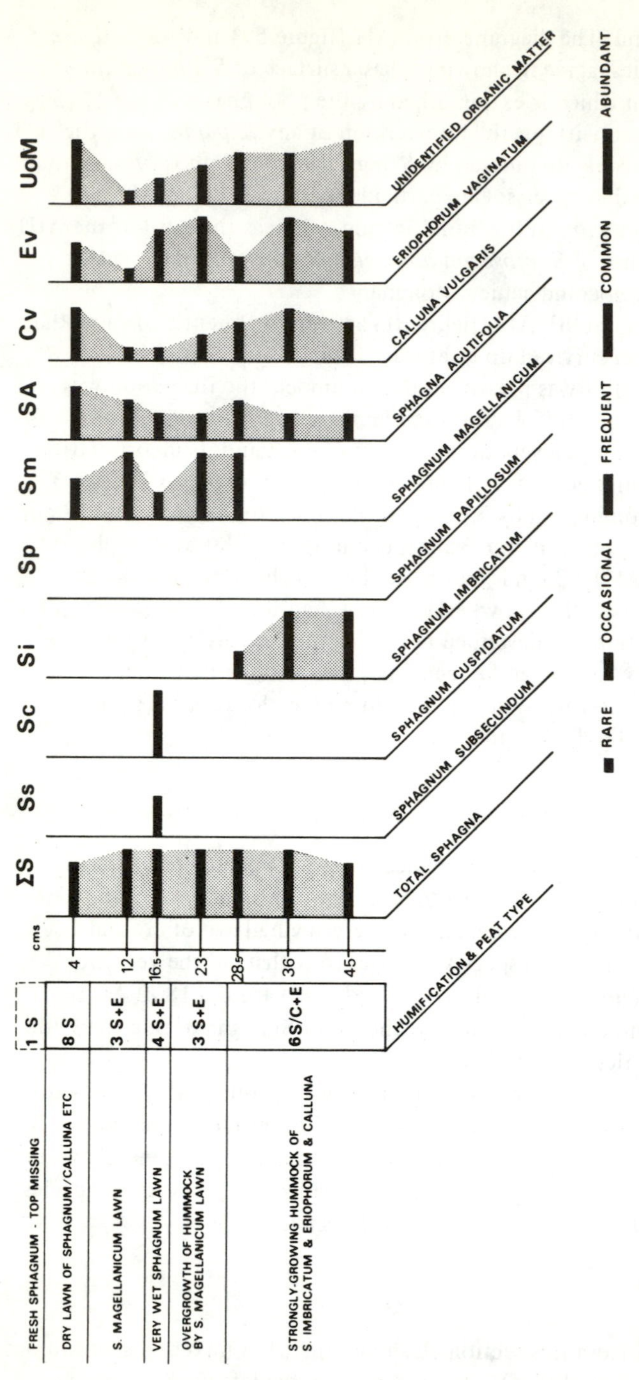

Figure 62. Macrofossil diagram from monolith BI1.

begins with a hummock dominated by *S. imbricatum* but this is interrupted by an occurrence of *S. magellanicum* at 70 cm (= 1360 AD), one of the rare occasions when the two species intermingled. The overgrowth of the hummock sees the end of *S. imbricatum* and its replacement by a lawn of mixed species, in which *S. papillosum* is quite prominent, which is in turn transgressed by the pool which covers most of section CI. *S. imbricatum* lasts until about 1580 AD in this monolith; in monolith CI2 its behaviour is rather different (figure 64). At the base of monolith 2, around 600 AD, *S. imbricatum* is growing in a rather wet situation with *S. cuspidatum,* and then disappears with the first pool-mud formation at about 700 AD. This pool-complex with three pool-muds was obviously an unhealthy habitat for *S. imbricatum,* as indeed it was for most other Sphagna, to the extent that the sample from 86 cm was simply a humified 'streak' of greasy mud within which no moss remains were identifiable to species. As drier conditions returned *S. papillosum,* Sphagna Acutifolia and *S. magellanicum* are all represented, but as the hummock spreading over the former hollow becomes increasingly humified so we see the late return of *S. imbricatum* – the 45 cm count dates from 1685 AD and the species was still forming the hummock when sealed in by the over-burden from the 1800 AD ditch.

Monolith CI3, only 30 cm to the left of CI2, is of interest in showing that *S. imbricatum* was present at the edge of the pool ecomplex for at least part of the time when it was missing from monolith 2. At 72 cm in profile CI3 (figure 65) it is abundant (dated to 1200 AD) and still present in the wet lawn at 64 cm (1325 AD) and although its place is taken by *S. papillosum* and *S. magellanicum* for a while it does reappear at 45 cm, the same level as in monolith 2. This sort of detail shows the danger of arguing too far from a single monolith, and also demonstrates once again that *S. imbricatum,* where it does survive beyond the general increase in wetness of the period 1300-1450, does so in hummock situations from whence it may recolonize small areas when they become suitable. Finally in monolith 3 it is a wet complex of *S. cuspidatum* and *S. magellanicum* which ends the profile in 1800 AD.

5.4.8 Section DI

The one monolith from this section shows a two-fold division very similar to that of monolith HI4, which is, of course, also similar stratigraphically. The *Sphagnum imbricatum*-Acutifolia-*Calluna* community of the lower part of the diagram (figure 66) is at first remarkably free of *Eriophorum vaginatum* which invades at about 62 cm to become abundant as *S. imbricatum* declines. The striking stratigraphic change to lighter-coloured peat of wetter origin is accompanied by an equally striking change to *S. magellanicum* dominance after a phase of mixed Sphagna associated with the 1450 AD pool. With few variations (for example, at 21.5 cm) this *S. magellanicum* community continues up to

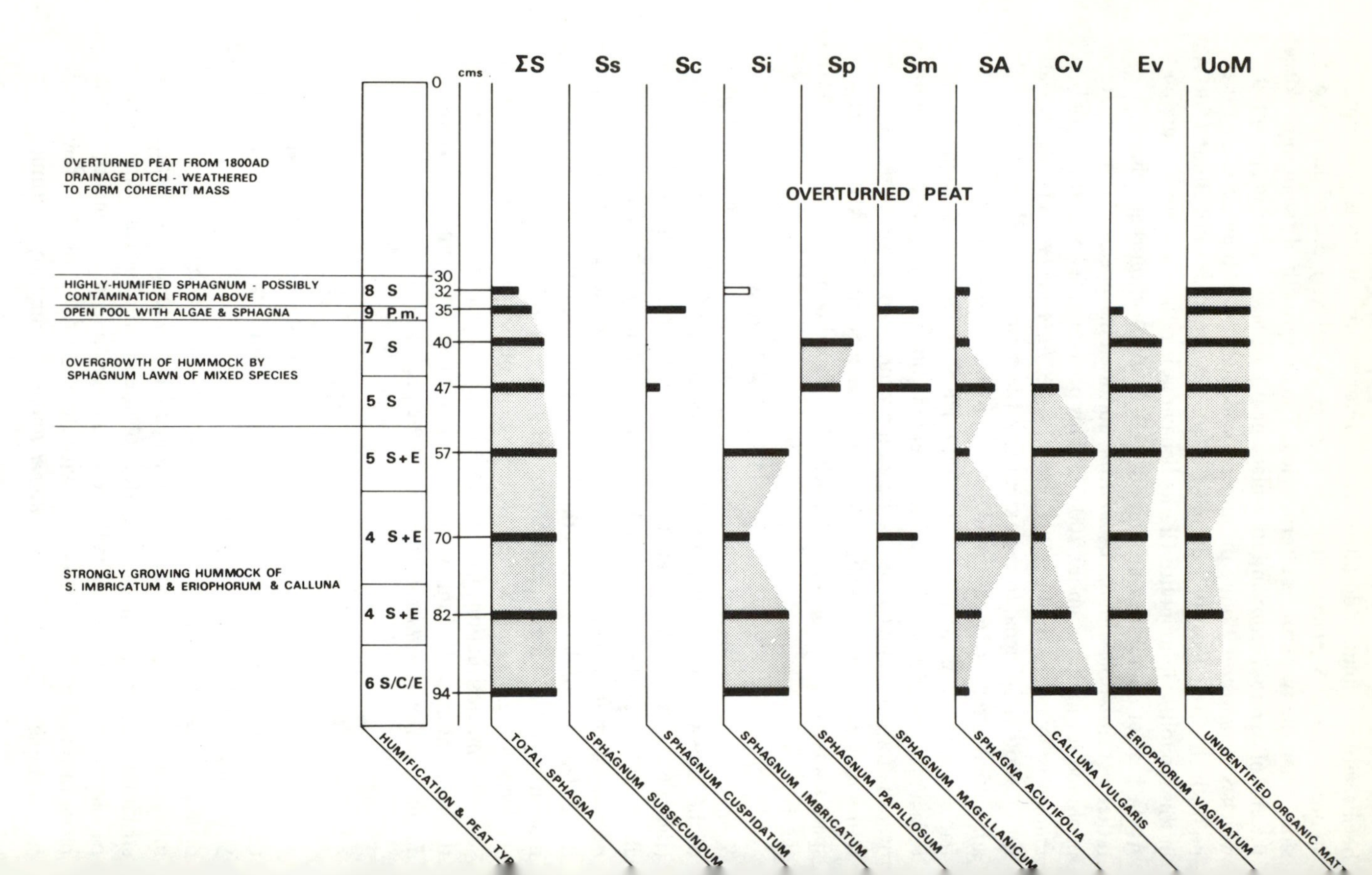

ΣS
Ss
Sc
Si
Sp
Sm
SA
Cv
Ev
UoM
cms
0
30
32
35
40
47
57
70
82
94
OVERTURNED PEAT
OVERTURNED PEAT FROM 1800AD DRAINAGE DITCH - WEATHERED TO FORM COHERENT MASS
HIGHLY-HUMIFIED SPHAGNUM - POSSIBLY CONTAMINATION FROM ABOVE
OPEN POOL WITH ALGAE & SPHAGNA
OVERGROWTH OF HUMMOCK BY SPHAGNUM LAWN OF MIXED SPECIES
STRONGLY GROWING HUMMOCK OF S. IMBRICATUM & ERIOPHORUM & CALLUNA
8 S
9 P.m.
7 S
5 S
5 S+E
4 S+E
4 S+E
6 S/C/E
HUMIFICATION & PEAT TYP
TOTAL SPHAGNA
SPHAGNUM SUBSECUNDUM
SPHAGNUM CUSPIDATUM
SPHAGNUM IMBRICATUM
SPHAGNUM PAPILLOSUM
SPHAGNUM MAGELLANICUM
SPHAGNA ACUTIFOLIA
CALLUNA VULGARIS
ERIOPHORUM VAGINATUM
UNIDENTIFIED ORGANIC MATT

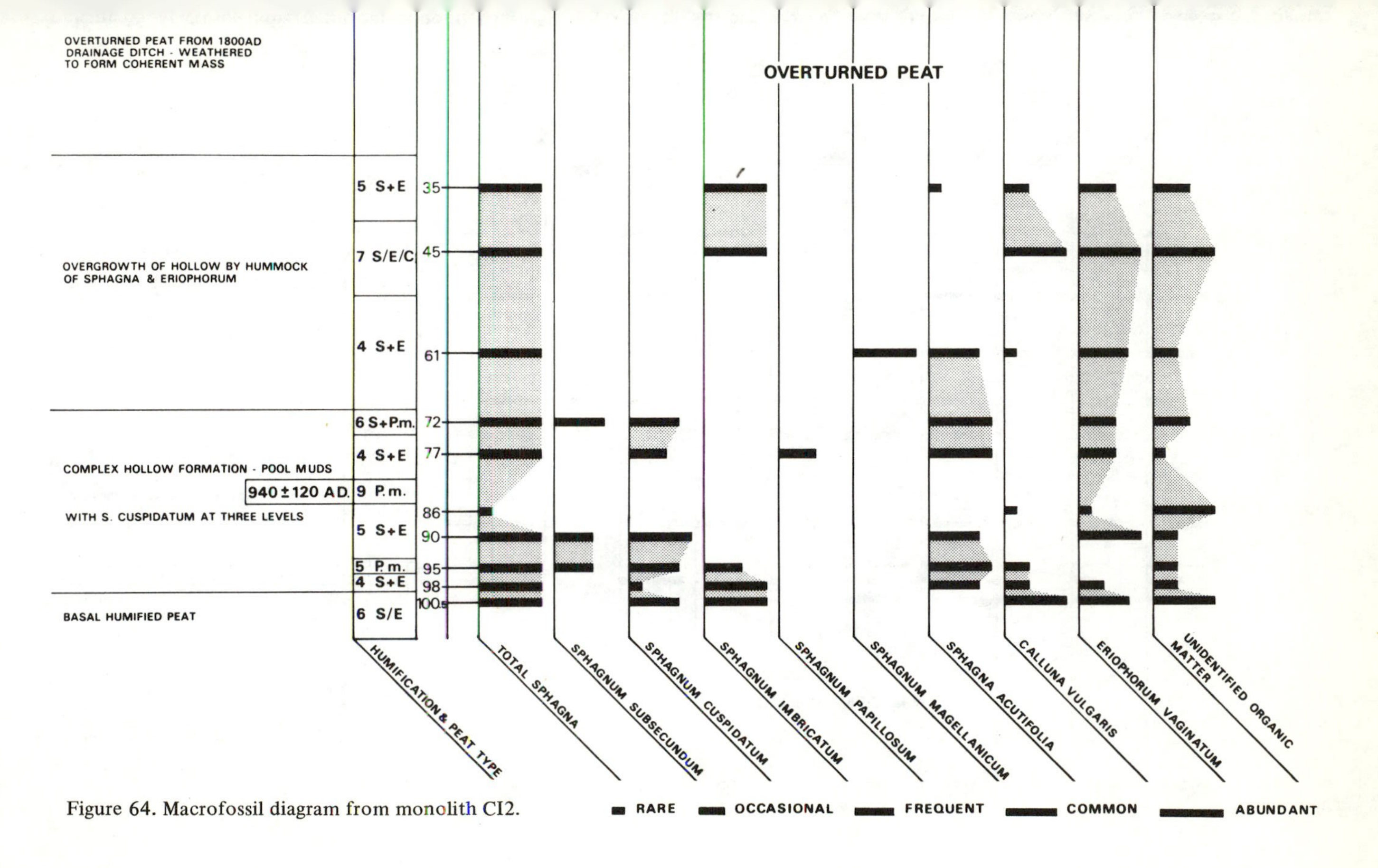

Figure 64. Macrofossil diagram from monolith CI2.

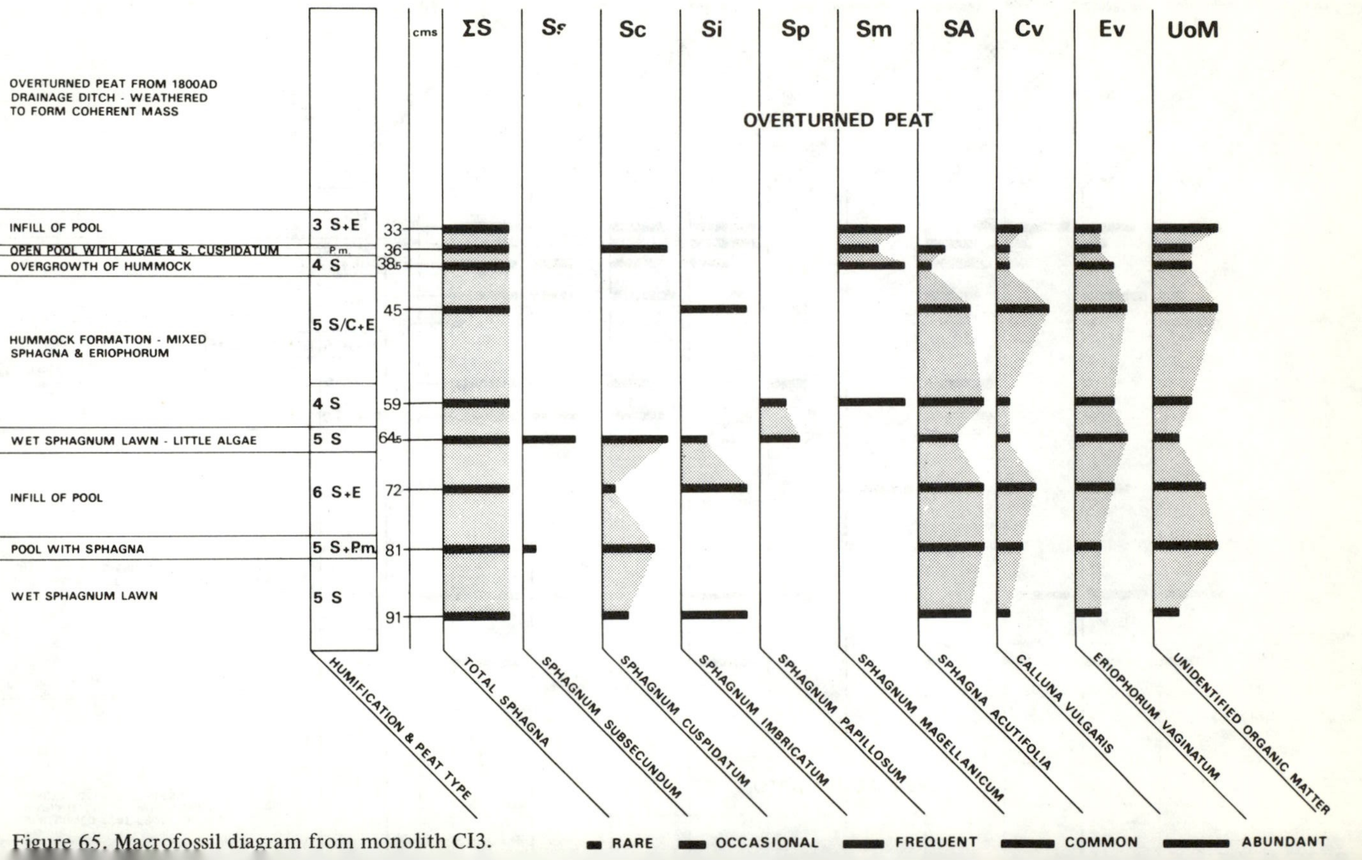

Figure 65. Macrofossil diagram from monolith CI3.

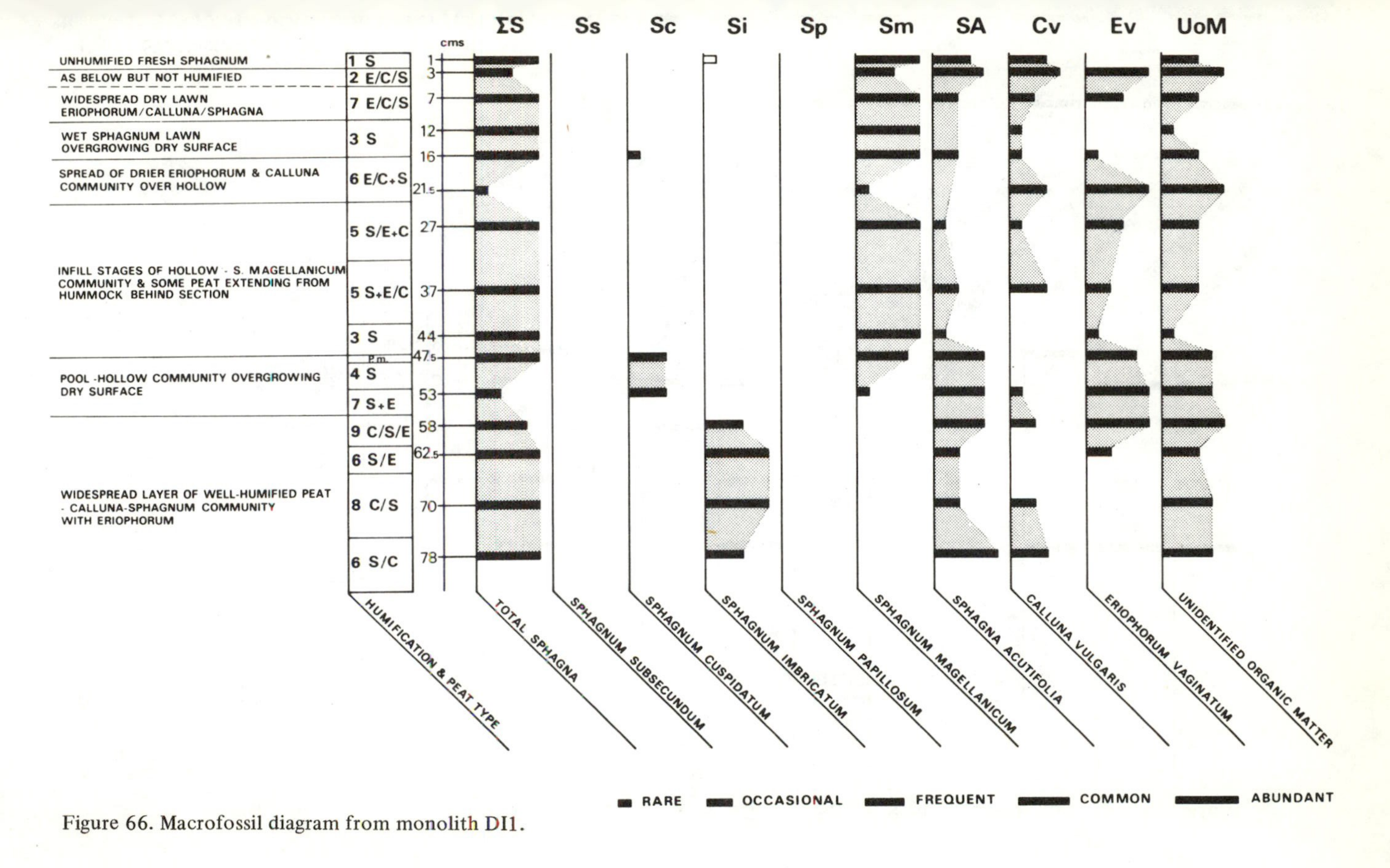

Figure 66. Macrofossil diagram from monolith DI1.

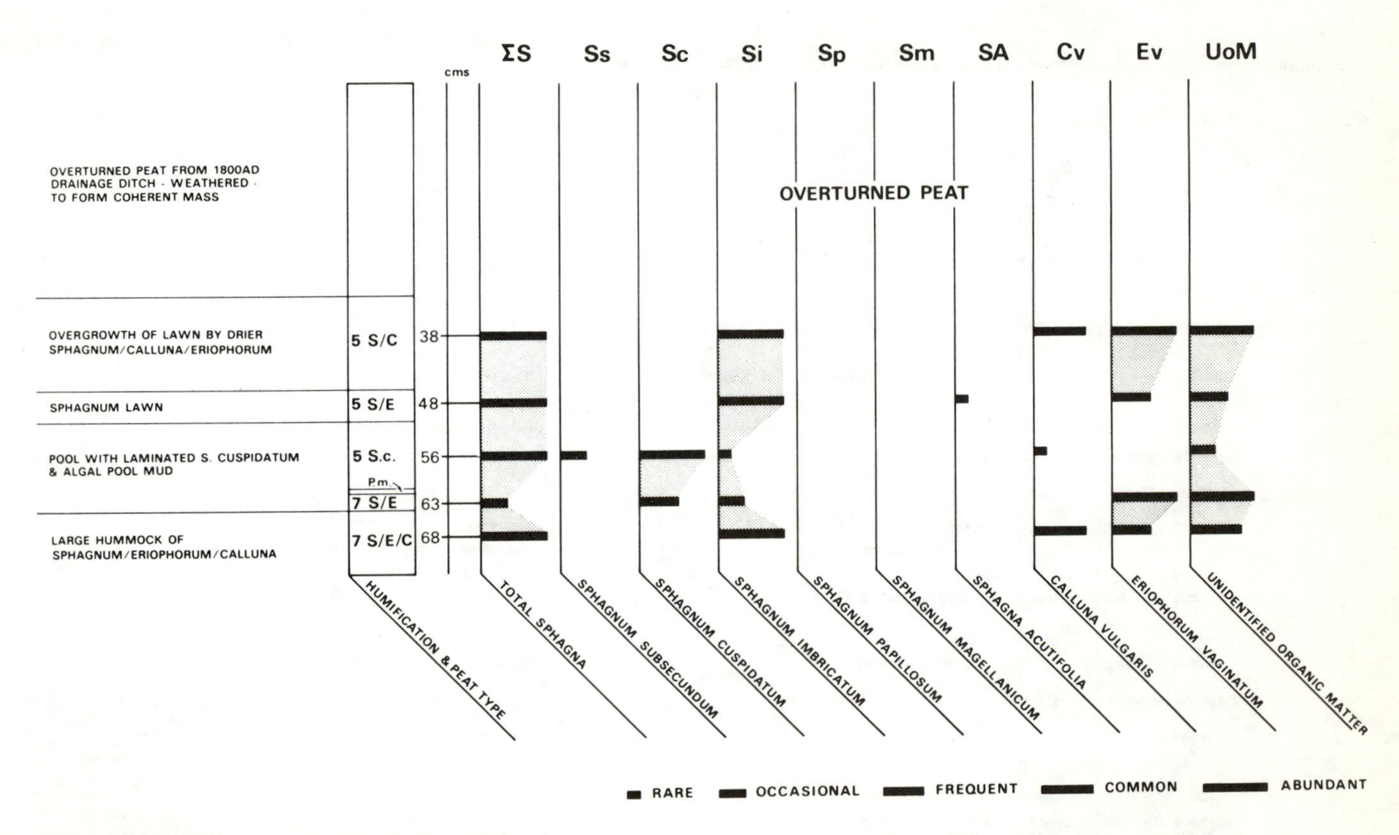

Figure 67. Macrofossil diagram from monolith GI1.

the 1950's surface – five centuries of peat accumulation due to almost a single species and with no hint of cyclic processes in either these results or the peat stratigraphy.

5.4.9 Sections GI and GII

Monolith GI1 displays a simple macrofossil profile (figure 67). The basal hummock is once again rich in *S. imbricatum,* to the exclusion of other *Sphagnum* species, together with *Calluna* and *Eriophorum.* The laminated pool peat above the hummock also contains *S. imbricatum,* albeit only at the lowest level at 56 cm, and the species apparently survives some centuries in this habitat or nearby (leaves and branches could easily have floated into the pool), to reassert itself in the *Sphagnum* lawn above the pool. There is no representation of *S. papillosum* or *S. magellanicum,* which accords with the previous evidence of floral composition of the whole bog at the time in question (up to 350 AD).

The monolith from section GII (figure 68) also begins with a *S. imbricatum* phase but in this case the species shows itself remarkably able to invade the wet lawn created after the 1050 AD pool layer which temporarily extinguishes *S. imbricatum* (54 cm count). The whole of the pool infill succession between 58-38 cm was very wet to judge from the humification and presence of *S. subsecundum* and *S. cuspidatum* and yet *S. imbricatum* became abundant at 1340 AD (45 cm), was present in the pool-mud of 1480 AD (38 cm), and in the final greasy pool-like layer around 1800 AD (25 cm). Almost certainly the reason for this pattern is to be found in the presence of a strongly growing hummock 20 cm to the right of the monolith (see figure 19) which would act as a nucleus of the species which could then spread into the hollow alongside whenever conditions favoured it.

Towards the top of the monolith (which was sealed in by overburden at 1800 AD), *S. papillosum* and *S. magellanicum* make their appearance in a characteristic sequential manner and had the peat here continued growth for a further 150 years it seems probable, based on the evidence of the other analyses, that *S. magellanicum* would have become dominant.

In both these monoliths one can relate little to cyclical processes; rather the *Sphagnum* succession seems conditioned by a series of jumps or shifts in the bog's hydrology leading to a pool, followed by a pool-infill sequence.

5.4.10 Summary of macrofossil evidence

From the 21 profiles analysed and presented above, a number of common features stand out as worthy of comment. Foremost amongst these is the behaviour of the three cymbifolian Sphagna, and the most striking phenomenon is the dominance of *Sphagnum imbricatum.* It is present in the lower part of every profile without exception and abundant in most. It cannot be found

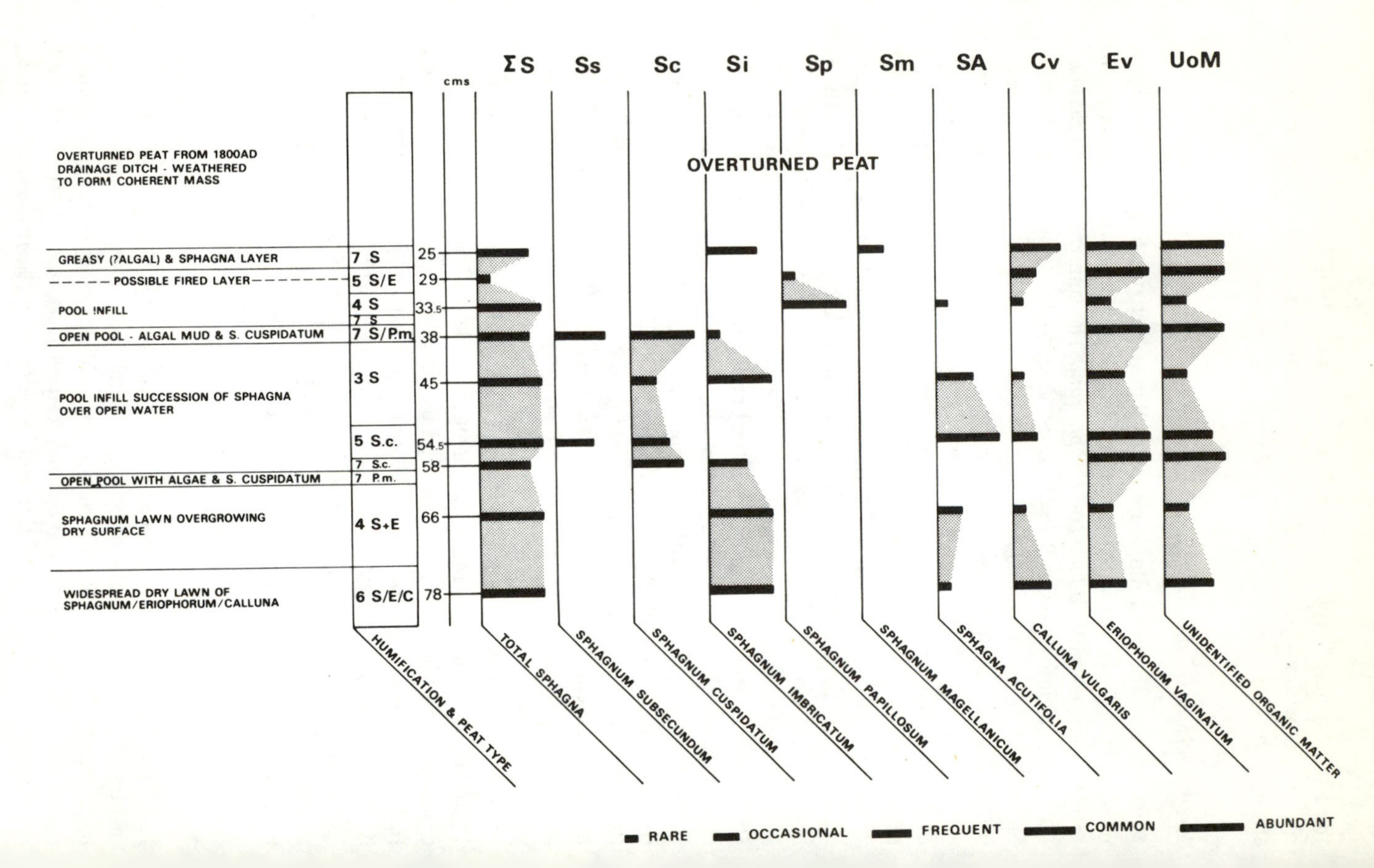

ΣS
Ss
Sc
Si
Sp
Sm
SA
Cv
Ev
UoM
cms
OVERTURNED PEAT
OVERTURNED PEAT FROM 1800AD DRAINAGE DITCH - WEATHERED TO FORM COHERENT MASS
GREASY (?ALGAL) & SPHAGNA LAYER
POSSIBLE FIRED LAYER
POOL INFILL
OPEN POOL - ALGAL MUD & S. CUSPIDATUM
POOL INFILL SUCCESSION OF SPHAGNA OVER OPEN WATER
OPEN POOL WITH ALGAE & S. CUSPIDATUM
SPHAGNUM LAWN OVERGROWING DRY SURFACE
WIDESPREAD DRY LAWN OF SPHAGNUM/ERIOPHORUM/CALLUNA
7 S
5 S/E
4 S
7 S
7 S/P.m
3 S
5 S.c.
7 S.c.
7 P.m.
4 S+E
6 S/E/C
25
29
33.5
38
45
54.5
58
66
78
HUMIFICATION & PEAT TYPE
TOTAL SPHAGNA
SPHAGNUM SUBSECUNDUM
SPHAGNUM CUSPIDATUM
SPHAGNUM IMBRICATUM
SPHAGNUM PAPILLOSUM
SPHAGNUM MAGELLANICUM
SPHAGNA ACUTIFOLIA
CALLUNA VULGARIS
ERIOPHORUM VAGINATUM
UNIDENTIFIED ORGANIC MATTER
RARE
OCCASIONAL
FREQUENT
COMMON
ABUNDANT

living on the present bog surface even though it was present as late as 1800 AD in hummock communities. Dates of its extinction vary from the Middle Ages up to 1800 AD and a major factor in this appears to be an expansion of wet lawn and pool areas. Its place has been taken by *S. magellanicum,* often preceded by a phase rich in *S. papillosum,* in both lawn and hummock situations, and with only one or two exceptions *S. imbricatum* and *S. magellanicum* do not occur together, and *S. magellanicum* is not found in the earlier parts of the profiles.

Sphagnum subsecundum and *cuspidatum* are, as expected, found in pool situations, the latter species in all pools and persisting as part of a mixed *Sphagnum* lawn after pool infill, whereas the former species is not found in all pools and there is a tendency (in accord with its known ecology) for it to be a component of supposed deeper pools, that is pools with a thick mud bottom.

The Sphagna Acutifolia group is difficult to summarize in that it undoubtedly includes ecologically different species such as *S. tenellum* and *S. rubellum* but there is a tendency observable in a number of profiles for this group to be important in hummock situations, especially once *S. imbricatum* has disappeared, and therefore the species present is very probably *S. rubellum.* On the other hand high values of the group in pool and lawn situations are not that unusual and probably indicate a fair proportion of individual stems of *S. tenellum,* possibly with some *S. recurvum.*

The total Sphagna column is notable for its high values throughout most of the profiles, the few lower values being associated with humified hummock situations, especially those rich in *Eriophorum vaginatum,* and with well-developed pool-muds. *Calluna* is also highly correlated with drier hummock situations and is absent from some profiles for a number of samples in succession if conditions are wet (e.g. HI4, 8 and 9). *E. vaginatum,* on the other hand, can apparently tolerate quite wet conditions and is often found from wet lawn situations in various profiles. Finally one may note the correspondence of the Unidentified Organic Matter curve and the humification values of the peat; thus the 'UOM' curve acts as a useful 'second opinion'.

5.5 CLIMATIC RECORDS

5.5.1 General considerations

The broad climatic conditions under which peat forms above the mineral-soil water limit, that is ombrogenous peat or tertiary peat (Moore & Bellamy 1974), are not as well defined or researched as may at first appear. Moore & Bellamy (1974) are vague on this point, mixing up their discussion of the 'climatic template' with a new classification of European mire types; so that neither their map of mire types in Europe nor the climatic diagrams give any clear indications of, for example, the extent to which a long cold winter, with much water from snowmelt, may compensate for a short dry summer. There is, in fact, very little literature on this point of what exactly the climatic requirements of raised bog growth are, beyond vague references to the 40 inch isohyet and an 'oceanic climate' (Godwin 1975 and Walker 1970) but quite obviously conditions in the humid west of Britain, from Somerset northwards, are such that ombrogenous peat formation can take place. Given the very minimum conditions then, a bog nearer to the boundary of such conditions, whatever they may be in terms of rainfall, evapotranspiration, rain-days per annum, etc., will be more 'sensitive' to climatic change than one well within the minimum-conditions boundary. Bolton Fell Moss seems to be such a 'sensitive' mire both on the evidence of its stratigraphy already described and its position on maps of average annual rainfall (Ordnance Survey 1949, 1967) where it can be seen to lie on the periphery of the Carlisle area 'rain-shadow', and on its position on Green's (1964) map of potential water deficit.

Spring and summer conditions may reasonably be expected to be more important in determining plant growth but peat accumulation will also depend upon the depth to the anaerobic sulphide layer (Clymo 1965), which will also depend on autumn and winter rainfall and temperatures. As a broad generalisation though, it may be said that low evapotranspiration in summer, due to high rainfall and/or low temperatures, will favour the production of unhumified *Sphagnum* peats and may cause an increase in the number and area of pools; cold winters, maintaining a high water table, will tend to accentuate this trend. Such conditions of increasing wetness have been shown to lead to decreasing production of *Calluna* and *Eriophorum* (Forrest & Smith 1975), a trend in accord with the macrofossil results from Bolton Fell Moss. Conversely, low water levels do reduce *Sphagnum* growth (Clymo 1973) so that hot, dry summers would lead to more humified peat (by lowering the sulphide layer), with a lesser *Sphagnum* content, greater amounts of *Calluna* and possibly a tendency to a 'reduction of surface variance' (Clymo 1973), that is, towards the formation of flattish dry surfaces.

The sections that follow give an outline of the temperature trends over the period represented by the main stratigraphic sections of Bolton Fell Moss, followed by a consideration of rainfall variations, the two then being combined in Lamb's 'High Summer Wetness' and 'Winter Severity' indices (1966). After brief consideration of some other local climatic evidence, some correlations of the climatic records with the peat and macrofossil stratigraphy are presented.

5.5.2 Temperature records

Direct temperature observations are available back to the year 1659 AD (Manley 1959, 1974) and using 'proxy data' Lamb (1965a, 1977a) has extended this record back to around 900 AD. Ladurie (1972) has presented similar data for continental Europe, particularly the Alps. Further evidence, often in striking agreement with that of Lamb and Manley, has come from oxygen isotope profiles of the Greenland ice sheet, both long and medium term (e.g. Johnsen *et al* 1972) and relatively short-term, covering the period 1200-2000 AD (Dansgaard *et al* 1975).

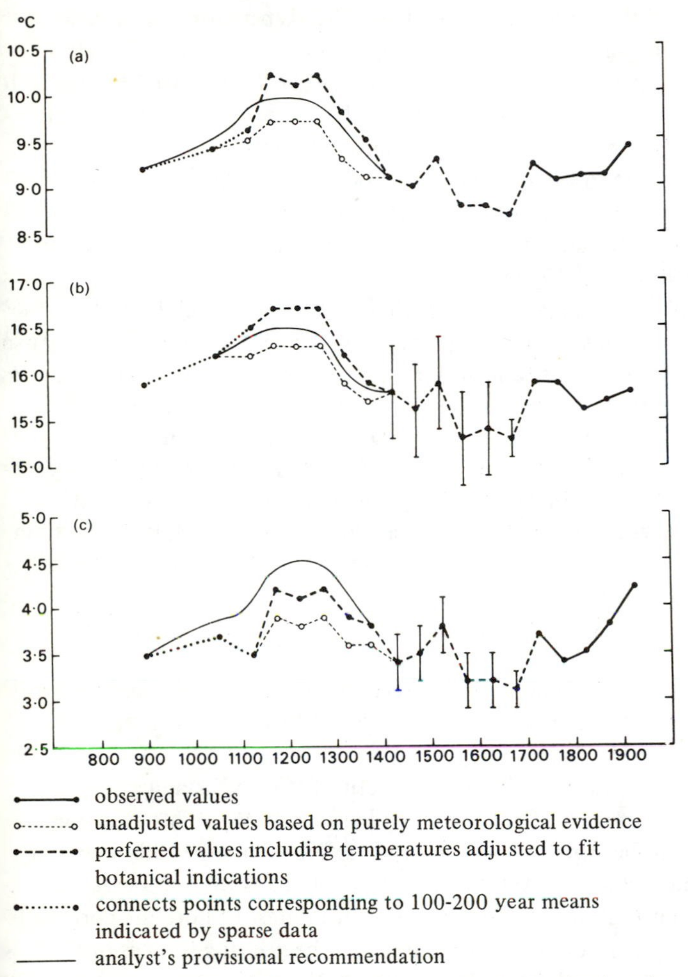

Prevailing temperatures (°C) in the lowlands of central England: estimated 50-year means since AD 1100 and longer-term averages before that. (a) whole year; (b) high summer (July and August); (c) winter (December, January and February) (from Lamb 1965).

Figure 69. Temperature trends in England since 900 AD, 50-year averages (from Lamb 1977a).

Whilst there are some arguments over detail – Ladurie (1972, p.265) disagrees with many other authors in the timing of the Little Ice Age – the main outline of temperature changes since Roman times are clear, and for those earlier parts of Bolton Fell Moss stratigraphy (e.g. section GI) a fair assessment of climate can now be made.

Our present climatic period, the 'cool, moist sub-Atlantic' (relative to the post-glacial climatic optimum, Lamb (1966)) began sometime around 900-500 BC and the date of the pool in section GI, 880 BC, is perhaps a manifestation of this. Lamb (1966) notes that annual mean temperatures around 500 BC would be something like 1 °C lower than today's (9.4 °C in Central England) and that this decline from a level estimated at 1 °C above present at around 1000 BC would have meant a 'great reduction of evaporation and wetter surfaces than before even in summer' (Lamb 1966, p.173). This is, of course, the period of the *'Grenzhorizont',* the main recurrence surface of north-west European bogs and the validity of this change is generally accepted (Godwin 1975, Frenzel 1966, West 1977).

The course of the temperature trends thereafter, until about 1000 AD or so, is rather uncertain. In his recent compilation of all available data, Lamb (1977b points out that there appears to have been a return to warmer conditions during the Roman occupation of Britain (AD 43-410) which led to a generally good period of climate between 400 and 1200 AD, though not without some setbacks. The graphs, reproduced from Lamb (1977a) as figure 69, show the general trends from AD 900. The Medieval Optimum in temperature is clearly shown and the 50 year means of this diagram are of a sensible sort of length in palaeobotanical terms. The deterioration which is known to have set in post-1300 AD, with disastrous harvests in 1315, 1316 and 1317 (Ladurie 1972), reached its peak in the Little Ice Age proper between about 1550 and 1850 AD, though a short period around 1530 is thought to have recovered almost up to today's values. The data presented recently by Dansgaard *et al* (1975), strikingly vindicates these trends, especially the 1300-1500 AD deterioration doubted by Ladurie (1972).

From 1670 AD onwards we have the benefit of Manley's standardization of the temperature trends in Central England (1959, 1966, 1974). The great glacier advances in the Alps and elsewhere are reflected in especially low temperatures around 1680-1700 AD (Manley 1966), and a period of increasing surface wetness on Bolton Fell Moss, and it is of interest to note the correspondence of the upper pool level of the bog and the glacial advances of the 1780-1800 AD period (Ladurie 1972, pp.200-207).

5.5.3 Rainfall records

These records are less reliable than the temperature trends from the earlier periods and only become reliable from 1727 onwards. The results presented by Lamb (1966, 1977a) show pronounced dryness of the summers of the High Middle Ages (1100-1300 AD) as shown in figure 70. During the Little Ice Age summer rainfall is thought to have fluctuated markedly, and figure 71 (from Lamb 1977b) shows the excess values for July and August, obviously vital months for plant growth, and the increasing tendency to wet springs and autumns. Allied to the coolness of the summers from 1300-1850 AD this graph shows long periods particularly favourable to the growth of Sphagna.

The records can confidently be applied to Bolton Fell Moss because of the close correspondence between the local record from Carlisle with the national record from 1727 to 1964 kindly supplied by the Meteorological Office. The local data, from Spital Cemetery, Carlisle, are continuous from 1864 and are

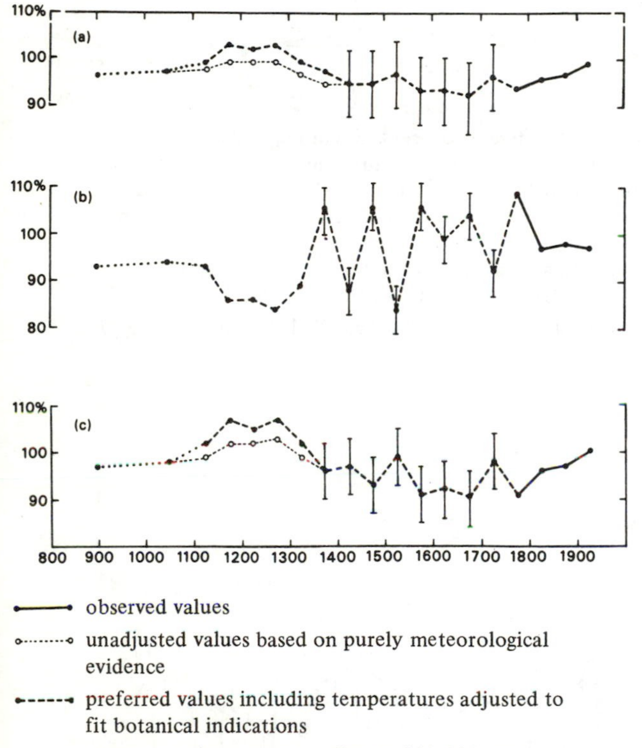

Average rainfall over England and Wales (as percentage of the 1916-50 means): estimated 50-year means since AD 1100 and longer-term averages before that. (a) whole year; (b) high summer (July and August); (c) the other 10 months (from Lamb 1965).

Figure 70. Rainfall trends in England since 900 AD (from Lamb 1977a).

plotted on figure 72, both as annual totals and as 10 year running means. Also plotted on figure 73 are the Central England figures, and a good measure of correspondence of the means can be seen, particularly between 1864 and 1884, and 1908 to 1956. However the coincidence is not nearly so good from 1890-1910. Statistical tests for this data are not without their problems – a X^2 hypothesis test fails to exploit the numerical nature of the data (Wonnacott &

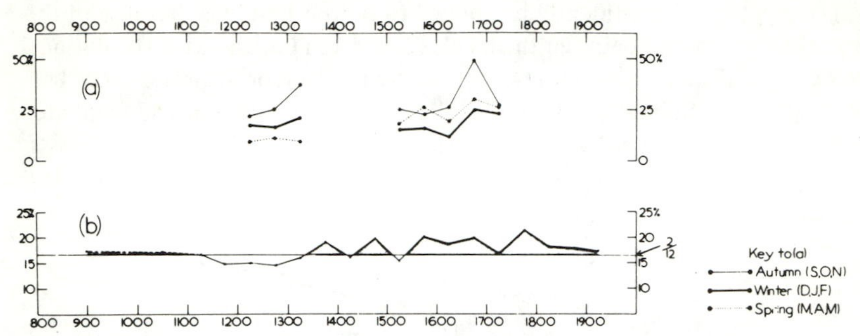

Seasonal rainfall distribution in different half-centuries in England and Wales.
(a) Percentage of months described as wet in the autumns, winters and springs (snowy months have not been included as wet, because the equivalent rainfall is nearly always below normal). Data were insufficient before 1200 and between 1350 and 1500.
(b) Percentage of the year's total rainfall falling in high summer (July and August). The black area indicates more than two twelfths of the year's total.

Figure 71. Seasonal rainfall distribution in England and Wales (from Lamb 1977b).

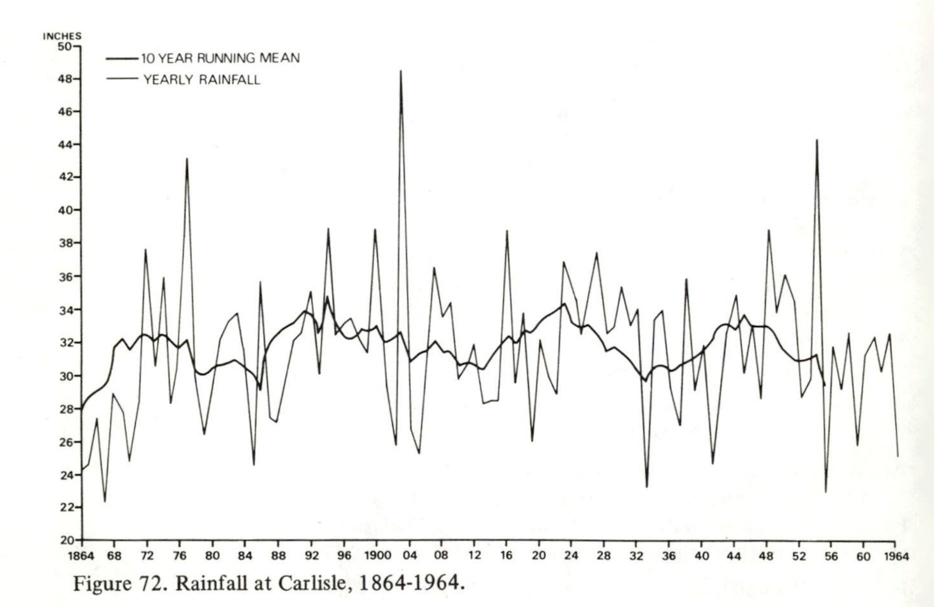

Figure 72. Rainfall at Carlisle, 1864-1964.

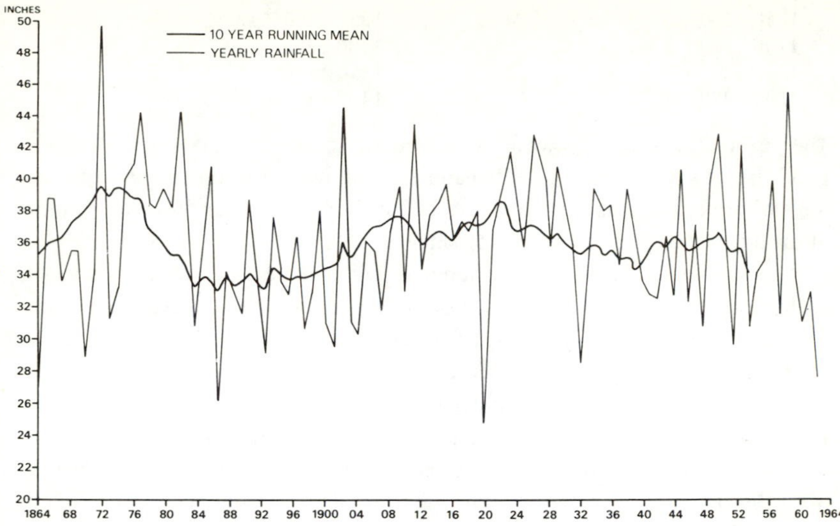

Figure 73. Rainfall over England and Wales, 1864-1964.

Wonnacott 1972) and serial autocorrelation problems make it difficult to apply a regression model to such data. A simple 't test' was used, therefore, to test the significance of apparently wetter and drier spells in both Carlisle and Central England. This showed that three wet periods, 1868-77, 1918-27 and 1942-51 were significantly different with at least a 99 % level of probability (most at 99.9 %) from the drier periods, 1878-87 and 1930-39 in *both* the England and Wales record and Carlisle.* Parry (1975) working on climatic change and marginal agriculture in the Lammermuir Hills, has also noted the correspondence of both temperature and rainfall trends in lowland Britain and upland southern Scotland. The application of pre-1864 rainfall records from Southern Britain to northern Cumbria would therefore seem justifiable.

Two further local sources of information on rainfall must be noted. Firstly, from the Medieval period, there is evidence of excessive drought between 1250 and 1300 AD in the settlement of the Norman Tower of Carlisle Cathedral. Messrs Cooper, Higgins & Partners, consulting structural engineers, found evidence of a 30 cm settlement of the tower and concluded that this had occurred shortly after building and certainly before 1292 AD due to a 'change in the subsoil structure . . . reducing the pore water pressure and hence the bearing capacity' (R.M. Higgins, personal communication). The engineers related this event to the literary evidence for droughts contained in Britton (1937) which, totalled in 50 year periods give the following:

* I am indebted to Kelvyn Jones for these computations and for advice on the statistical tests to be used.

1050-1100	3	1250-1300	12
1100-1150	4	1300-1350	5
1150-1200	3	1350-1400	6
1200-1250	9	1400-1450	1

The years 1252 and 1253 were exceptionally severe drought years (Britton 1937, Brooks & Glasspoole 1928) and even allowing for under-recording of the earlier periods it is clear that the 1200's had a significantly greater number of droughts than periods before or since.

The second local source of evidence is from the peat bog itself by way of iodine analyses. The results from monolith AI1 (hollow) and AI2 (hummock) are shown together on figure 74, together with a number of date-points taken from the pollen correlations of section 5.3.3.

These results both agree, and disagree, with what is known of rainfall variations from other sources, and as explained in section 4.4.2, the method must be developed further before it can be reliably used in peat. The curves from the two monoliths show a fair measure of correspondence at first sight,

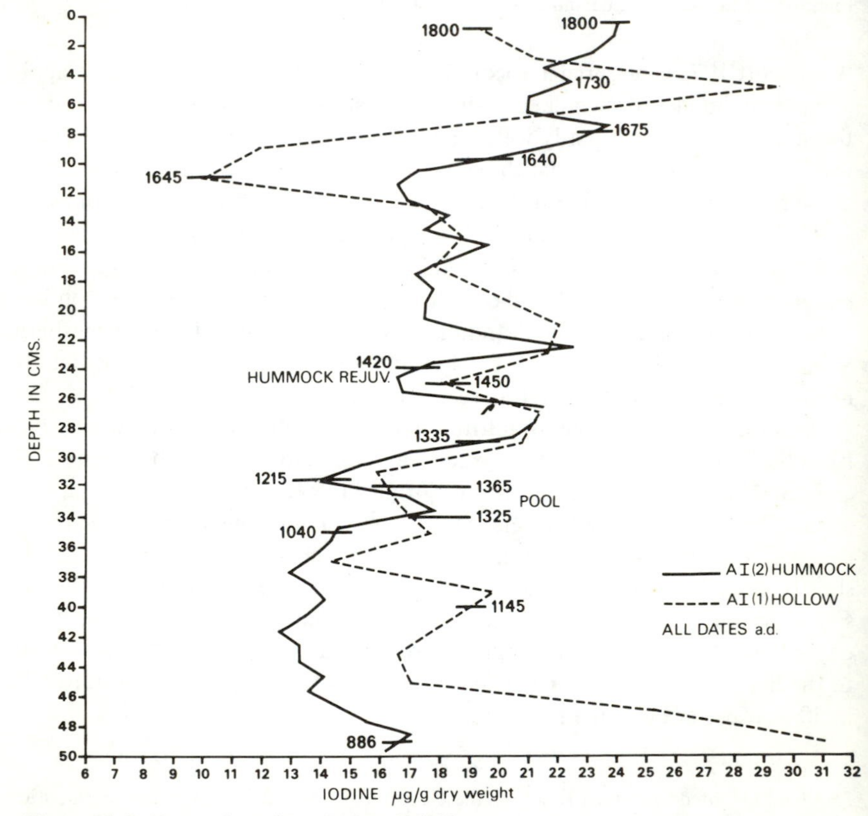

Figure 74. Iodine analyses from Bolton Fell Moss.

but in the lower half of the curves much of this correspondence is false, in that the dates may be decades or centuries apart. Although one could argue that both curves show the 1200's to be relatively less oceanic than the 1300's and that the late 1400's are here relatively wet, agreeing with, for example, Lamb's evidence, the extreme values in the AI1 curve at 49 cm, 11 cm and 5 cm (1025 AD, 1645 AD and 1730 AD) do not fit with the stratigraphy or the macrofossil trends and indeed the 1730 AD value – a very high one indicating an oceanic wet period – conflicts directly with the Meteorological Office 1727-1964 record in that 1729 had 100.6 % of the annual average (1916-1950 = 35.6 inches), 1730 = 81.2 %, 1731 = 64.3 %, 1732 = 82.3 %, 1734 = 114.9 %, etc. In other words if the period was abnormal it was abnormal for its lack of rain rather than a surfeit of it. There is also great disagreement of the curves around 1640 AD. Despite these contrary results, which can only be resolved by much more detailed work on peat/iodine relationships along the lines of Pennington & Lishman's (1971) work on lake sediments, there does seem to be a basic division of both graphs with a 'drier' Medieval phase and a 'wetter' phase post-1300 AD.

A final piece of evidence from the Bolton Fell Moss area to the account of the great bog-burst of Solway Flow which took place on the night of 16 December 1771, and covered over 200 acres to a depth of several feet (Walker 1772). There had been very heavy rain ('. . . as had not been known for at least two hundred years' Walker (1772, p.124)) but even before this Solway Moss had been extremely wet – '. . . so much of a quagmire, that, in most places it was hardly safe for any thing heavier than a sportsman to venture upon it, even in the driest summers' (Walker 1772). There are further references to its wet state and estimates of its depth, etc. The date of this bog-burst almost coincides with the formation of the upper pools at Bolton Fell Moss and the higher rainfall is borne out by the actual figures for 1750-1800 AD (see also figure 71 for summer excess rainfall).

5.5.4 Correlations of climate and peat stratigraphy

The combined effect of the two parameters, temperature and rainfall, is best brought out by figure 75, taken from Lamb (1977b). The parts of both graphs which refer to Bolton Fell Moss (the far left hand side; BFM longitude 2°50′W) show a pattern which agrees very well with the humification and character of the peat, especially when taken with the July-August rainfall graph for England and Wales (figure 71). From these and the tables in Lamb (1966, pp.217-9), it is clear that there was an excess of wet summers during the periods 900-1050 AD, 1320-1500 and 1560-1850. The latter two periods were also colder in both winter and summer (figure 69), the later period being the 'Little Ice Age' proper, separated from the earlier decline by a resurgence of warmer, drier climate around 1530. Although much more detail, and some changes of detail,

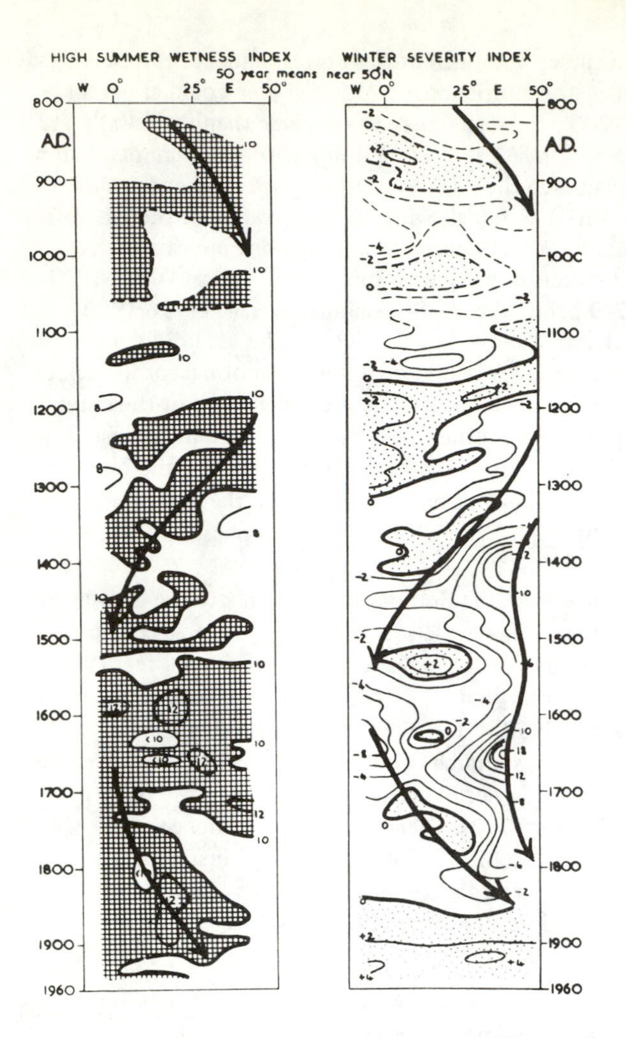

50-year mean values of the Lamb summer and winter indices at different longitudes in Europe near 50°N between England and Russia from AD 800 to 1960.
Hatched areas indicate excess of wet over dry months in high summer (July and August).
Stippled areas mark more very mild than severe winter months (December, January and February).
Schematic arrows make it easier to follow the movement of the greatest anomalies during times of change.
In the periods of climatic standstill – i.e. when some persistent character of the atmospheric circulation must be presumed (AD 1050-1200, 1550-1700, and after 1900) – all longitudes tended to fare alike as regards prevalence of dryness or wetness, warmth or cold.

Figure 75. Summer wetness and winter severity indices (from Lamb 1977b).

have been incorporated in Lamb's 1977 compendium, these diagrams, originall drawn in the mid-1960's, have not been altered and give the authoritiative climatological view.

While it is clear from the foregoing results of stratigraphic and macrofossil analyses that there are a great many correlations between these climatic trends and events on the bog – the dryness of the late 1200's (monoliths HI4 and DI1 for example) and the subsequent deterioration – it was thought that the relationships would be made even clearer if certain stratigraphic events were extracted from the record of all the monoliths analysed and dated, and that these events might then be graphed.

Consideration was given to using both 'dry' and 'wet' stratigraphic events but the 'dry' markers were not systematically used for obvious reasons – they could result from pool infilling in a constantly wet climate and they were 'phases' rather than markers, and so could not be given a single date. Instead they were taken into the final picture when the 'surface wetness' curve was being drawn – figure 76. What could be used systematically were the shifts from a dry phase to a wet phase, 'dry' and 'wet' being used to describe the bog surface condition and not necessarily rainfall since, of course, low temperature could either mask a decrease in rainfall or accentuate an increase.

These wet 'phase-shifts' were identified and dated from the macrofossil and pollen results and are of two kinds. The first and lesser kind is a shift from a dry community to a wet lawn situation, such as the overgrowth of the hummock extension in monolith HI5a. The second kind is a phase-shift to a definite pool formation, identified by its algal mud and presence of *Sphagnum subsecundum* and *cuspidatum,* which obviously denotes a more severe change in conditions. Where a wet lawn precedes a pool phase by only a centimetre or so only the pool date has been used, but where there is a sizeable gap, representing 50 years or more, then both phase-shifts are registered (e.g. HI4 shifts at 1360 and 1425 AD).

The results (figure 76) show a striking coincidence to Lamb's climatic curves, with the wettest periods occurring between 900-1100 AD, 1320-1485 and 1745-1800. The middle pool of most stratigraphic sections stands out particularly, starting at 1425 AD in seven monoliths. The secondary periods of wet lawn shifts precede all the pool phases and show up well in the Little Ice Age proper.

The curve of surface wetness is drawn up on the evidence of these shifts but also includes consideration of the general stratigraphy of the several monoliths used, especially with regard to the dry periods of 90-c.600 AD, the 800's and the 1200's, and the early 1500's. It is not therefore an exact 'quantitative' curve, and probably it never could be, bearing in mind the heterogeneous nature of the stratigraphic evidence, but it is a considered 'analyst's opinion', similar to the gloss on Lamb's curves (figures 69 and 70), and based on evidence such as the temporary drying out of monolith DI1 between 1510 and 1610 AD, hummock extension phases in section HI, etc. The 1200's are more clearly marked as a period of greater dryness based on evidence such as the ^{14}C-dated dry phase of 1270 in monolith HI4, and the 800's are noted as a lesser dry phase during which the 'widespread humified surface' of the HI monoliths formed, followed by the pools of 900-1085. The period from 1150-1200 should probably be seen as an interruption to a rising curve rather than regarding the 1130 peak as a separate entity – there is good evidence from sections such as AI that the 1100's were dry, at least the early 1100's.

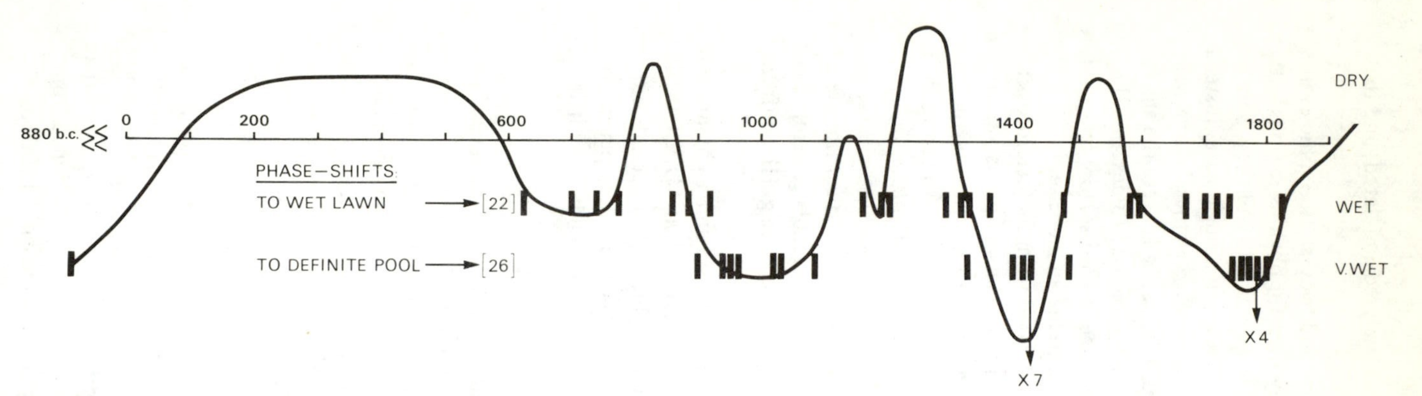

Figure 76. Surface wetness curve of Bolton Fell Moss since 90 A.D.
N.B. Curve is generalized and includes data other than wet phase-shifts shown – see text for details. Relative height of 'dry' peaks reflects area and vegetation/humification of hummocks.

The Medieval and Late-Medieval part of this surface wetness curve is even more interesting if compared with the details supplied by Lamb (1965b, p.30) on extremely dry and wet decades:

a) *Summers with over 120 % of the July and August rainfall average of England and Wales, 1916-50*

1310 – 9)	
1360 – 9)	
1450 – 9)	4-6 in each case
1490 – 9)	
1560 – 9)	

b) *Summers with under 70 % of the July and August rainfall average of England and Wales, 1916-50*

1130 – 9	3-4
1230 – 9	4
1240 – 9	4-6
1270 – 9	3-4
1290 – 9	4
1300 – 9	3-4
1330 – 9	3
1500 – 9	3-5

Beyond 800 AD the surface wetness curve has few correlative pieces of evidence, though there are some interesting new reports in Lamb (1977b, Chapter 17). These show a tendency to a warmer drier climate than now in the 300-400 AD period, but cool summers between 500-700 AD – a good fit with the evidence presented here. The 880 BC pool of section GI stands alone as the oldest stratum examined and may be taken to represent a change akin to the *Grenzhorizont*, though of course further dating evidence is necessary to confirm this.

We see, therefore, in both the graphs of surface wetness and of climate, that the peat at Bolton Fell Moss has been very sensitive to climatic changes over the last two millenia or more, reacting swiftly to changes in temperature and/or rainfall in forming pools or else broad dry surfaces which are revealed to us by their characteristic stratigraphy and macrofossil content.

6. DISCUSSION

The results detailed in section 5 give clear proof of climatic control of the peat stratigraphy at Bolton Fell Moss. No other factor or group of factors, biotic or abiotic, can so fully explain the changes in field stratigraphy and macrofossil results, though of course such factors as burning or natural water tracks may have influenced parts of particular profiles. In many peat sections (BI, DI, for example) there is no evidence of cyclical processes at all, even though the peat has taken several centuries to accumulate, and where there is, on the face of it, an alternation of hummocks and hollows (CI, GII and HI) these can be fairly precisely related to independently-known climatic changes. This being so, and in the light of the many doubts expressed by the authorities discussed in section 2, it is necessary to abandon the theory of cyclic regeneration and erect a new hypothesis.

6.1 POOL FORMATION

This phenomenon is an obvious starting point for discussion in that the Bolton Fell Moss pools are 'datum lines' – they can often be traced over a wide area and, give or take a few decades in some cases, they are contemporaneous features. The way in which they have formed across humified and unhumified peat alike in section HI, at 1425 AD in response to climatic change, supports Walker & Walker's ideas (1961, pp.182-4) of a bog surface reacting in the same way all over rather than lending weight to the ingenious arguments of Boatman & Tomlinson (1973, 1977). Both of these papers contain a great deal of data on the Silver Flowe bogs and in particular on the origin of pools. They come down against the cyclic theory (1977 paper, p.544), and note that once formed, pools tend to be long-lived features but that 'difficulty arises in understanding how a pool originates' (p.545) and they also note their rather contrary results from an examination of cores from below pools. These results showed *Sphagnum papillosum* to be important in a number of places beneath existing pools, as well as a tendency for pools to originate over peat which was already in a semi-aquatic condition. This is, however, based on only 15 cores from under two pools and from three short transects; a much fuller picture can obviously be seen in the side-wall of a ditch. Their tentative explanation of pool formation – that it owes a lot to the poor performance of *S. papillosum* and *S. cuspidatum* in slight hollows so that more vigorous surrounding areas grow up around them, the hollow eventually becoming a pool – is at variance with the Bolton Fell Moss data. This is not to say that such a mechanism as Boatman & Tomlinson postulate cannot exist and their data on hummock growth versus *Sphagnum* lawn growth – showing a tendency for hummocks to rise faster – is interesting in this regard. However, as is clear from the results of, say, section HI, where individual features can be followed over *centuries* of growth, any approach based on surface form and growth rates over even the 20-25 years achieved by Boatman & Tomlinson (1977, p.541 – using growth rings in the adventitious roots produced by *Calluna* as it is overgrown by *Sphagnum*) is bound to get at only a very small part of the picture.

A further idea to explain pool formation, put forward by Tallis (1973) and supported by Dickinson (1975), is that of drainage impedance by humified and even 'algae-encrusted' peat. While the whole field of hydraulic conductivity in peat is very involved and somewhat controversial (Ingram 1967, Rycroft, Williams & Ingram 1975a, 1975b) it is generally true that unhumified or slightly humified peat will transmit water at a much faster rate than humified peat, wherein water appears to be physically bound in some sort of colloidal complex and where also the pore space may be easily blocked (Rycroft *et al* 1975b). This being so, and practical manifestations of it such as water seepage from a peat face above a recurrence surface are commonly observed, then it seems plausible that a widespread humified surface or 'retardation layer' (originally used to mean retardation of peat accumulation, Godwin 1954) could cut down infiltration sufficiently to allow the water table to rise and flood the bog. There are, however, logical difficulties here – how do humified layers ever form to the sort of thickness common in bogs, 50 cm or more of H8-9 *Calluna-Sphagnum* peat? The idea also is at variance with evidence such as that of the stratigraphy of section HI where the middle and upper pools appear indiscriminately across humified and unhumified peat alike. However the idea cannot be dismissed and could well contribute to major wet-phase shifts – there could for example, be a threshold involving a combination of climatic change and percentage area of drier surface on each particular bog.

It can be seen from the above discussion, from the literature surveyed in section 2 and from the evidence presented in section 5, that pool formation and infill is a crucial process to understand in any comprehensive hypothesis of bog growth. They are not, as has occasionally been suggested, an incidental phase to the main regeneration process involving *Sphagnum* succession because, as shown in the results from Bolton Fell Moss and from the Walker's Irish work, the *Sphagnum* succession often changes *because of* pool formation.

It may also be said that while pools are hydroclimatic features the hollows which they occupy are not and so the essential problem is to explain the formation of hollows. This argument may be countered on three points. Firstly, one may simply point to the impossibility of there being a perfectly flat mire surface of any extent in nature – variation is the rule. Secondly not all pools are in definite hollows. The pools of section HI are good examples of the variety of pool form and size which clearly depended upon the particular form of the surface when the wet phase-shift in the climate occurred. To paraphrase Boatman & Armstrong (1968) hollows may be looked upon simply as 'non-hummock places' and so the origin and size of depressions is bound up with the initiation and differential growth of hummocks – and, one might add, variations of level too small even to merit the term 'hummock'. This leads on to the third point – that the chance scatter of small areas of differential growth, for whatever reason, will inevitably produce hollows. The hollows and pools may even widen and deepen themselves by chemical means as postulated by Sjors (1946), or simply by being sub-optimal habitats for *S. cuspidatum* which then dies out and so a semi-permanent 'algal pool' is established which becomes deeper by the growth of the surrounding lawns and hummocks (Boatman & Armstrong 1968, Boatman 1977). This leads on to consideration of the role of *Calluna* and other species and discussion of growth and accumulation rates, which are dealt with below (sections 6.3 and 6.4).

Finally one must review the ideas concerning pool formation, recurrence surfaces, and major and minor climatic changes. In attempting to reconcile and unify the variety of views and evidence one is reminded of the far-seeing paper of Conway (1948), and of Godwin & Conway (1939) – see sections 2.3 and 2.4. The quotation from Conway (1948) on page 24 is now seen to be truly prophetic in the light of the correlation of climate and stratigraphy at Bolton Fell Moss. Climates have now been proved to be unstable and one could suggest that we certainly do have to 'alter our interpretation of the observed vegetational mosaic which we call the regeneration complex'. Not only that but in view of the persistence of pool and hummock features between the recurrence surfaces, and then on up to the surface on Bolton Fell Moss, I would suggest that we have to incorporate recurrence surfaces fully into a new hypothesis of bog growth. Granlund's (1932) last two recurrence surfaces –RYI at about 1200 AD and RYII at about 400 AD – are represented at Bolton Fell Moss, and they are part of the normal build-up of peat. The datings differ, but then Granlund was not using C^{14} dating and since then the dates have been shown to lie in a scatter around Granlund's original points – Overbeck (1963) gives around 600 AD for RYII and 600 BC for RYIII but claims another RY at 0 AD. From the Bolton Fell Moss dates for wet phase-shifts and the graph of surface wetness one could make a case for RYIII at 880 BC, RYII at 625 AD and RYI at 1200 AD (in AI1) – but then there are 26 phase-shifts to pool conditions and 22 shifts to wet-lawn. Instead of regarding recurrence surfaces with an attitude of 'exceptionalism' this data – and the confusion of dates for various 'RYs' from all over north-west Europe – suggests that we should instead look upon recurrence surfaces and their widespread pool layers as an integral part of the normal growth of a raised bog, and regeneration complexes (sometimes, recalling Tregaron, complexes not of the bog centre) simply as an active peat-building phase following climatic rejuvenation.

A final point regarding bog pools is that of size. Obviously large bog ponds such as described by Osvald (1923) and Casparie (1972) have special reasons attaching to their formation, such as the virtual coalescence of bogs which Casparie describes, but the smaller kind, ranging perhaps from 20 cm to 2-3 m across, have exercised the minds of various authors, some of whom seem to have been unsure of whether such pools 'belonged' to a Regeneration Complex, to a Teichkomplex *sensu* Osvald 1923, or were residual from a recent flooding of the bog. Walker (1961) discusses this at some length (p.32) and with some difficulty. He attempts to differentiate between large and small pools (3-6 m compared with 1 m pools) and argues that the large pools in Irish bogs which were 'much more common' than the small pools and occurred 'at a few, fairly distinct, levels', did not 'belong to the usually accepted regeneration complex cycle . . . they did not necessarily lie above hummock peat, and that, when they did so, they also transgressed the hollows between'.(Walker 1961, p.32). This argument is dropped from the 1961 joint paper altogether, and the pools are described simply as they appear. It is clear that the size argument is recognized as having dubious validity so much depending upon the hummock density and hummock size of the flooded surface, and the depth of flooding.

6.2 SUCCESSION OF SPHAGNA

There is little published work on *Sphagnum* succession comparable to the results from Bolton Fell Moss. Walker & Walker's (1961) results are from a 'fluctuating flat' type of stratigraphy not involving *S. cuspidatum*, and from an adjacent rather flattened hummock and hollow involving *S. cuspidatum* and *S. subsecundum*. Both sets of results include *S. imbricatum* with no hint of its demise and both are undated. Whilst the first profile, from Fallahogy bog, is not of great interest, the second and third profiles (Walker & Walker 1961, figures 2 and 4) display a very similar type of succession to that which occurred in, say HI4 and HI6, or CI3 and CI2, on Bolton Fell Moss, with hygrophilous Sphagna overgrowing a mature surface, first in a hollow area, and then gaining some representation on the hummock sides which became rejuvenated with *S. imbricatum* and continued as a hummock. It is noticeable that *S. magellanicum* and *S. papillosum* play very minor roles indeed.

Tolonen (1966 and 1971), though mentioning *Sphagnum* remains throughout and producing some tables of abundance, prefers to rely on rhizopod diagrams to indicate surface conditions and does not graph *Sphagnum* remains. Casparie's work (1969, 1972), particularly from his site Emmen – 17, shows long periods of dominance by single *Sphagnum* species (1969, figures 6, 8 and 9) with sudden changes between species – for example profile 17B which changes from a centuries-long dominance of *S. rubellum-Calluna* to *S. papillosum-cuspidatum* and then to *S. imbricatum.* This is very similar to the types of changes at Bolton Fell Moss, though with different species (Casparie's profile is all pre-100 AD and *S. imbricatum* was still extant) and, like the author, Casparie can see no cyclic succession here – and this over a period of 2000 years as at Bolton Fell Moss, so that the records combined cover the last 4000 years.

More recently Aaby & Tauber (1974), Aaby (1976) and van Geel (1976) have shed more light on peat profile changes, *Sphagnum* successions and climatic changes. The first paper has a single macrofossil profile showing something of an alternation in abundance of *S. imbricatum* and *S. papillosum* after a long period dominated by *S. cuspidatum.* In the second paper no further information is given on macrofossils but a firm statistically-tested relationship is established between peat-growth and climate. In van Geel's exhaustive analyses of a single profile from a Dutch bog, dating from before 4300 BC until 300 AD, the macrofossils show first a long period of domination by Sphagna Acutifolia (1400 BC-900 BC) with no cyclical changes, then a fairly rapid change through a *S. papillosum-cuspidatum* phase to a bog dominated by *S. imbricatum* (825BC-300 AD) with virtually no other Sphagna. Peat formation is shown to be intimately related to climatic change with several interrelated forms of evidence, some never used before such as newly-identified fungal remains, some of which indicate wet conditions, some dry (van Geel 1976, figure 8, p.21). Although climatic cycles are recorded there is no evidence of cycles involving Sphagna.

In the light of these studies the results of the macrofossil analyses from Bolton Fell Moss are not out-of-line; indeed they tend, if anything, to show more variability which could be interpreted in a cyclical fashion were it not for the evidence of climatic change. The main line of development, from a *S. imbricatum*-dominated community to one equally dominated by *S. magellanicum,* by way of a *S. cuspidatum-papillosum* phase, is represented by 16 of the 21 monoliths analysed. The secondary line of development with *S. imbricatum* continuing in some form until at least 1800 AD, is shown from only four monoliths and in each case the survival of *S. imbricatum* was due to a hummock habitat. In every such case *S. imbricatum* was accompanied by *S. magellanicum* and one may speculate that, given time, such hummocks would have come to be dominated by *S. magellanicum* (see macrofossil diagrams AI2 and GII1, for example). There is no 'conflict' with the cyclic regeneration theory here in that

S. magellanicum can form hummocks just as well as *S. imbricatum,* as was recognised by Osvald (1923) and in his succession diagram *S. magellanicum* is depicted as more important than *S. imbricatum,* which only comes into the picture when hummocks are rejuvenated (figure 4).

The question of the extinction of *S. imbricatum* is a vexed one, as has already been indicated in section 2. Dickson (1973) summarized the evidence and the theories, pointing out that the climatic explanation depended on bog surfaces drying out (Godwin & Conway 1939, Godwin 1956 – also supported by Tallis 1964) but that *S. imbricatum* also occurs in central Europe and regions with dry continental summers and survived Green's (1968) desiccation experiment better than any other species. The evidence from Bolton Fell Moss is for the species to survive in a robust form in the driest and most humified of peat (HI9 basal peat, HI4 at 1270 AD) but for it then to abruptly disappear on the flooding of the surface. This is backed up by its survival on hummocks such as HI6 and its spread over the wet lawn of central HI during a slightly drier climatic period (macrofossil results, figures 55 and 57). One must also consider the affect of human interference with bog surfaces by drainage and burning (Pigott & Pigott 1963, Pearsall 1956). The first of these has had little effect on *S. imbricatum* at Bolton Fell Moss (see macrofossil diagram AI2, figure 62) as one would expect if it were surviving on drier hummocks, and there is no real evidence for frequent fires. The explanation involving changing trophic status of raised bogs (Green 1968), is not considered relevant in this instance since the close sampling shows it to be flourishing in the oligotrophic hummock top, probably with *S. rubellum* and *Calluna,* and then disappearing in the presumably more trophic environment of the pool and wet hollow above the old dry surface. The question of phenotypic plasticity between the hummock form and the lax form of *S. imbricatum* has already been mentioned, and it seems that in this case the species had become 'locked in' to a dry growth form and could not perform the rapid ecad change induced experimentally by Green (1968). It is therefore suggested that the extinction of *S. imbricatum* at Bolton Fell Moss was climatically induced, and due not to dryness but to excessive wetness around 1300-1400 AD, followed by a further wet-shift around 1780-1800 AD, possibly helped by competition from *Sphagnum papillosum* and *S. magellanicum.* This explanation would be in accord with the water-level preferences of the various species as reported by Ratcliffe & Walker (1958, figure 3 and table 3), and by Green (1968). The same goes for the other *Sphagnum* species, *S. subsecundum* and *S. cuspidatum,* which were found in just the sort of water-level situation reported by Godwin & Conway (1939) and Ratcliffe & Walker (1958). What may be a little surprising is the amount of *S. cuspidatum* present in the lawns of post-Medieval times (see macrofossil results for HI4, HI9, etc.). Such lawns were obviously very wet, in keeping with the climate (surface wetness curve, figure 76) and the present surface of the uncut area of Bolton Fell Moss is not a great deal drier – the dominant *S. magellani-*

cum has a water level only a few centimetres below the surface and is accompanied by much *S. tenellum* and frequent *S. cuspidatum* in the hollows. As can be seen from the stratigraphic diagrams this modern surface represents a phase of just-infilled pools from the 1780-1800 AD phase-shift.

6.3 THE ROLES OF *CALLUNA* AND *ERIOPHORUM VAGINATUM*

These two species are found to be correlated with dry phases in the stratigraphy of the bog, *Calluna* quite markedly so. *Eriophorum vaginatum* is known to colonize fairly wet lawn situations (Osvald 1923), and to do well in areas where the water level is high in spring but falls in summer (Wein 1973). The ecology of both species is quite well known, Gimingham (1972) containing most of the information available on *Calluna* – much of it relating to dry heathland sites where more work has been done rather than bog sites (e.g. Chapman *et al* 1975a, b). Wein's (1973) Biological Flora account is a good source of information on *Eriophorum,* much work on productivity having been done by Forrest (1971) who has also studied the productivity of the combined *Eriophorum-Calluna-Sphagnum* system.

These studies have shown, amongst other features, that *Calluna* in bog situations is 'potentially immortal' because it roots adventitiously as it is overgrown by *Sphagnum* (Forrest 1971), and that *Eriophorum* tussocks are very hardy indeed, growing at very cold temperatures and resisting fire well (Wein 1973). The author has observed the latter behaviour after a fire on Bolton Fell Moss; the *Eriophorum* tussock bases were the only vegetation to survive and thrived thereafter, helped no doubt by nutrients from the ash – potassium has been found to be the main limiting factor in growth (Goodman & Perkins 1968).

The recognition of the clear difference between *Calluna* growth in an active raised bog and on dry heathland is of fundamental importance. Watt's original (1947) paper uses the example of the regeneration complex, based upon Osvald (1923) and upon Godwin & Conway (1939), and unfortunately seems to assume the same fate for *Calluna* as he had observed in the Breckland in his work on grassland and heathland (1940) – that is that the species would undergo pioneer, building, mature and degenerate phases. While this is true of not only *Calluna* but also bracken, *Festuca ovina, Agrostis* and other species (Watt 1947), it is emphatically not so with *Calluna* on bogs (Forrest 1971, Forrest & Smith 1975, Boatman & Tomlinson 1977). This could explain the lack of degenerating hummocks seen in the field by various authors (for example, Ratcliffe & Walker 1958) and on Bolton Fell Moss and quite obviously it removes a major plank in the cyclic regeneration theory. A similar process of rejuvenation could well operate with *Eriophorum vaginatum* tussocks, Wein

(1973) mentioning tussocks remaining active for over 100 years and noting that 'tussocks may reach great ages although this has not been well quantified'. Almost certainly there is also some vegetative reproduction from plantlets, as can be done in the laboratory (Gore & Urquhart 1966).

This 'immortality' of these two vital components in bog vegetation is seen in both the stratigraphic and macrofossil diagrams from Bolton Fell Moss. It is indeed the rule for hummocks of both *Calluna* and *Eriophorum* to last for some centuries and this may be seen in, for example, sections DI and HI. In the latter section we see an example of a persistent hummock with a 'core' of *Eriophorum* and *Calluna* (figure 20), at 50 cm HS, and the hummock at around 550 cm HS actually migrates to the left of the section. Both structures survived for something in excess of 1,000 years. Where hummocks are overgrown, as happens in most sections, there is little sign in the field of autunomous degeneration and macrofossil analysis of the HI6 hummock showed that it had definitely been 'swamped' while in an active state. The HI3 hummock analysis (figure 51) was done in particular detail because of the sharp boundary between it and the peat above, and because of the black greasy nature of the hummock-top peat. However, as reported in section 5.4, the remains at 20 cm were not typical of a degenerating hummock top (*S. magellanicum* abundant, small rather than large *Calluna* twigs) and the same goes for the hummock in section BI. In other places hummock tops merge into a more widespread humified surface, which may at a later stage be overgrown by a swathe of less humified peat – section DI is a good example of this sort of sequence.

The initiation of hummocks is thought by Boatman & Armstrong (1968) to be due to *Calluna* and *Eriophorum* acting as 'scaffolding' for the weak-stemmed Sphagna. This idea receives strong support from records and analyses such as those from section HI, monoliths 3 and 6 (figures 52 and 56) where the small 'knots' of *Eriophorum* and *Calluna* surviving the general flooding act as sources of the plants for the hummocks above.

Similarly, existing hummocks are often rejuvenated by a rise in water level and so grow up in pace with the surrounding wet lawn. There are stratigraphic signs of this having happened in almost all sections and two which were analysed for macrofossils – AI2 and CI2 – show that in both cases the rejuvenated growth was of *Sphagnum imbricatum,* which is interesting confirmation of Osvald's (1923) observations on the Komosse bogs.

Two final points regarding hummocks have come out of the Bolton Fell Moss results. Firstly, the macrofossil analyses clearly show that hummocks, even seemingly 'pure' *Eriophorum* hummocks or tussocks, contain abundant *Sphagnum* remains (BI1, AI2, HI3, etc.). Careful excavation of hummocks by the author in the field confirm this, as do the observations of Boatman & Armstrong (1968) and Boatman & Tomlinson (1977). This is not perhaps always appreciated by those familiar mainly with the cotton-grass moors of the Pennines, where Sphagna, though abundant in the peat, are absent from the present surface, due probably to atmospheric pollution and burning. *Eriophorum vaginatum* is not, therefore, to be looked upon as intrusive, or not part of, the classic regeneration complex; Osvald (1923) specifically includes it. There are, of course, occasional 'knots' of dense *Eriophorum* leaf-bases but in terms of the whole volume of peat these are of little importance at Bolton Fell Moss. Secondly, Walkers' observations (1961) on hummock outgrowths and the colonisation of surrounding infilling hollows, are amply confirmed in detail in the Bolton Fell Moss peats – the oscillations of the 50 cm HS hummock in section HI are good examples. Not only that but as has been shown by the study of the hummocks in central HI – monoliths 3, 3a, 5a and 6 – the outgrowths of hummocks can be dated to a period of relatively warmer and/or drier climate within the Little Ice Age; that is to say, this is another example of climatic control on a fine scale, rather than simply a chance event.

6.4 GROWTH AND ACCUMULATION RATES

The relationship between the growth of live *Sphagnum* and the accumulation of *Sphagnum* peat is a very complex one. Modern production studies, preeminently the work of Clymo (1965, 1970, 1973 and Clymo & Reddaway 1974) give some clues as to species differences in peat production, especially when taken with experiments on breakdown of *Sphagnum* tissues (Clymo 1965), but Clymo himself is careful to warn against a too ready translation of figures from modern situations to fossil situations (Clymo 1965, p.757; 1970, p.45). It would be inappropriate here to embark on a long discussion of modern production studies, but the following points, of some bearing to the interpretation of historical data, may be noted:

a) *Sphagnum* breakdown rates vary and this breakdown mainly takes place above the anaerobic sulphide layer (Clymo 1965). In particular *S. papillosum* lost mass at only 50 % of the rate of loss of *S. cuspidatum* or of *S. acutifolium* in Clymo's experiments, and this could lead to unrepresentative fossil assemblages, as well as topographic differences. Against this must be set differential growth, which is greatest in pools (Clymo 1970) and the lesser depth to the sulphide layer in pool situations.

b) Species growth rates vary quite markedly in response to habitat conditions (Clymo 1970) and one of the factors inducing rapid growth seems to be the presence of *Calluna* (Clymo & Reddaway 1974). This ties in with Boatman & Armstrong's (1968) observations, and with the higher total production of both *Calluna* and *Sphagnum* observed when they grow together (Forrest & Smith 1975). A lowering of the water table causes the surface of a *Sphagnum* lawn to flatten out, due to increased evaporation from any plants which protrude, thereby producing a negative feedback mechanism (Clymo 1973), and growth is greatest or equal greatest in almost all species in pool conditions and least on hummocks (Clymo 1970).

c) Using the adventitious rooting of *Calluna* to date *Sphagnum* and to thus measure growth over longer periods than is practicable by direct observation, extending back up to 15 years. Boatman & Tomlinson (1977) have shown that *S. papillosum* lawns increase in height at a steady rate of 0.7-1.5 cm per year, and hummocks of *S. rubellum* had rapid bursts of growth of up to 3 cm per year, thus maintaining themselves above the lawn. They also consider that large hummocks do not grow as units but that some parts of them grow faster than others from time to time.

All these observations are echoed in the results from Bolton Fell Moss and additionally new information on peat accumulation rates has been obtained. In particular the way in which hummock growth can keep pace with infilling hollows is confirmed time and again from the stratigraphic sections, and when dated macrofossil profiles are considered (sections 5.3 and 5.4, wherein several accumulation rate figures are given) one is struck by the way in which hummocks have apparently grown at a very rapid rate. Of course, some of this must be explicable by the resistance to compaction of a *Sphagnum* hummock 'reinforced' by upright *Calluna* stems, as against weak-stemmed species such as *S. cuspidatum,* but by whatever means it is clearly possible for hummocks to 'keep their heads above water'! The phases of faster growth, and the differential growth of hummocks noted above are to be seen in many places in the Bolton Fell Moss sections. Hummock CI3 was rejuvenated several times, and the hummock near HI8 grew differentially and at a very rapid rate.

The flattening out of lawns under low water levels postulated by Clymo (1973) is seen especially well in section HI (the widespread dry lawn at the base, also the 1270 AD lawn phase) but it must be said that surface microtopography at times of high water level, such as 1300-1800 AD, is also not shown to be very great.

Peat accumulation rates show great variation from site to site and period to period when considered on a world scale encompassing the whole Flandrian stage, the last 10,000 years. On the scale of the last 2500 or so years, the 'post-*Grenz*' peats of north-west Europe exhibit much less variation. Walker (1961) gives a figure of 20 years/cm, a figure repeated with more backing from ^{14}C dates in Walker (1970). Hibbert & Switsur (1976) give rates of 19.5, 12.1, 12.0, 26.3 and 15.8 radiocarbon years per centimetre from their study of five raised bogs, and other authors give figures of the same order of magnitude, such as Aaby & Tauber (1974), where the average figure for Draved Mose is 22.7 years/cm and the extremes 12.5 years/cm to 62.5 years/cm.

The rates from Bolton Fell Moss fall into the same range, but in the uppermost peats there are some very rapid rates – doubtless partially due to lack of compaction, about which opinions vary, but also reflecting the very wet conditions there. The Age/Depth profiles of a number of profiles have been added

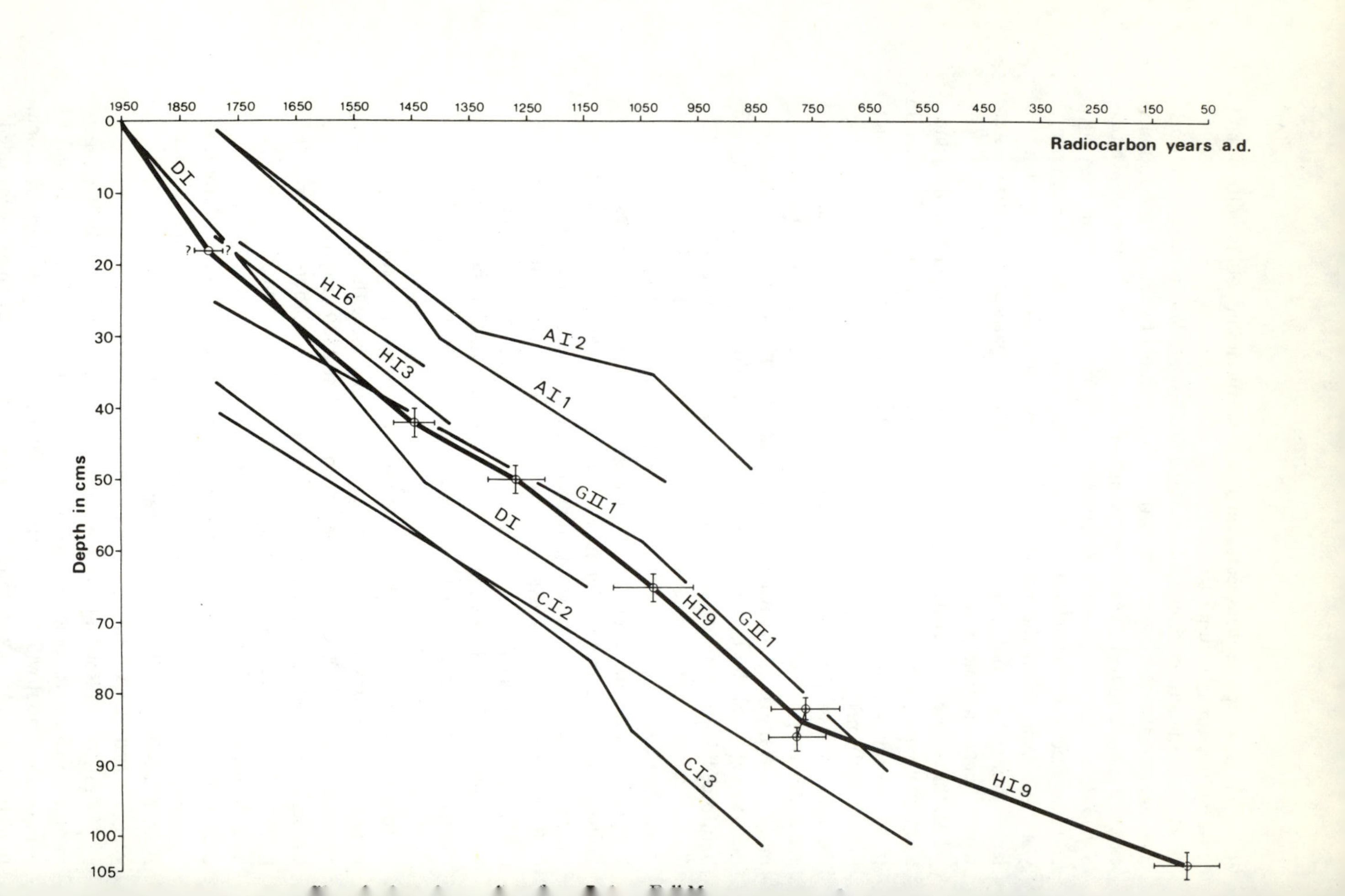
Radiocarbon years a.d.
1950
1850
1750
1650
1550
1450
1350
1250
1150
1050
950
850
750
650
550
450
350
250
150
50
Depth in cms
0
10
20
30
40
50
60
70
80
90
100
105
DI
HI6
AI2
HI3
AI1
GII1
DI
CI2
HI9
GII1
CI3
HI9

to that of HI9 to produce figure 77. The overall rate in monolith HI9, 17.75 years/cm (104.75 cm in 1860 years; 0.563 mm/yr) is in good agreement with other estimates; the extreme rate of 2.70 years/cm has a precedent in Turner's report of 3.3 years/cm from very wet peat from Tregaron Bog. It will be noticed that the Age/Depth profiles shown on figure 77 run almost parallel and that the inflexions in some curves are more or less contemporaneous. This is interesting in demonstrating that Bolton Fell Moss, though probably composed of two or more separate basins at an early stage in its development, was, in this ombrotrophic stage, growing as a unit and without too much of the spatial variations noted by Casparie (1972) in the raised bog complex east of Emmen. There is definite evidence of slower growth in more humified peat and hummock peats (lower part of HI9 before main humification change = RYII, and AI2 mid-hummock phase) but equally there is evidence of rapid spurts of hummock growth. The overall impression is one of the whole bog responding all over to changes in surface wetness, growing up in faster and slower phases, but without any suggestion of cyclic processes.

7. CONCLUSIONS AND FUTURE RESEARCH

This study set out to test the cyclic regeneration theory by palaeoecological means and in Popperian terms. One hopes that it may fairly be said that the hypothesis has been falsified. Osvald's succession diagrams, and those of Hansen and others, are valid as zonations of plant communities and in many cases as 'upgrade' successions, but there are no autogenic cycles, no regular 'downgrade' phases. The announcement of these conclusions in a paper read to the X INQUA Congress in Birmingham 1977 (Barber 1977) produced no disagreement with the proposition that 'the sequence hollow-hummock-hollow occurs only as a chance event', all the discussants agreeing that a new theory of bog growth was necessary.

From this demonstration of climatic control of even small-scale features of bog stratigraphy, control which has always been accepted on the larger scale of raised bog distribution and the medium scale of recurrence surfaces, a theory is now proposed which fully recognises the importance of wet and dry phases of climate – the Phasic* Theory of bog growth. This theory states that raised bog growth is controlled above all by climate, even down to the level of the relative areas of hummock and pool, and that the phase-shifts in peat growth are a result of climatic shifts. Threshold factors may cause the operation of the theory to differ from region to region and, to a lesser extent, from bog to bog, but the factors of hydrology and drainage, life-cycles of plants, pool-size, etc., are all subordinate to climate. The theory also unifies the formerly distinct phenomena of Recurrence Surfaces and Regeneration Complexes the latter being viewed as a natural consequence of the former, so that by the process of pool-infill and hummock-spread the bog tends to a drier state, until the next phase-shift.

The limits of upward bog growth require further study, which will prove difficult because of bog destruction by man, but it is obvious that there must be an upward limit, even allowing for lateral expansion and transgression over the lagg area. It may be that all bogs eventually 'flow' or burst, as did Solway

* *Phasic* appears in the Concise Oxford Dictionary (p.828, 1976 ed.) and has been used by Kershaw (1973, p.69) in describing phases of a cycle.

Moss and as Bolton Fell Moss may have done in the past to judge from aerial photographic evidence. Periodic flowage and subsidence could be a 'natural' event but as yet unrecognised because of the vast time-scale involved, relative to human lives.

One of the attributes of a theory is that it should predict events. At the moment the uncut part of Bolton Fell Moss is in a 'steady state'; there are no permanent pools and only isolated large hummocks, there is almost 100 % *Sphagnum* cover in this 'early mature' phase, the 1780-1800 AD pools having just been filled. The next stage, to judge from the climatic prediction of Dansgaard and his co-workers, could well be another wet phase-shift with quite small pools and low hummocks, beginning before the end of the century.

Further research at Bolton Fell Moss could involve closer work on the more or less distinct *Sphagnum* 'increments' in some hollows; some study of auto-compaction factors involving bulk density and penetrometer measurements; Catastrophe Theory modelling of peat growth in collaboration with mathematicians, further work on iodine fluctuations in mono-specific peat, and the dating of the recent peat using new methods such as ^{210}Pb determinations so as to relate present productivity to the peat formed during the last 200 years.* Now that the peat at this bog has shown itself to be such a sensitive recorder of past climates, the immediate plans involve extending the record back in time and exploiting the great advantage of the site, that of stratigraphy which is relatable to known climatic fluctuations and also accessible for study. There are a few sites in Europe, and virtually no others in Britain, which have these unique attributes and where the old geological dictum can be reversed and applied to an ecological problem: 'The past is the key to the present'.

* Peat samples from monoliths HI8 and HI9 have now been used in a palaeomagnetic study (Oldfield, Thompson & Barber 1978) and in a study of radioisotopic dating (Oldfield *et al* 1979).

8. REFERENCES

Aaby, B. (1976) Cyclic climatic variations in climate over the past 5,500 yr. reflected in raised bogs. *Nature* 263: 281-284.

Aaby, B. & H. Tauber (1974) Rates of peat formation in relation to degree of humification and local environment, as shown by studies of a raised bog in Denmark. *Boreas* 4: 1-17.

Aiton, W. (1811) *Treatise on the Origin, Qualities and Cultivation of Moss-earth, with Directions for converting it into Manure.* Air.

Andersen, S.Th. & F. Bertelsen (1972) Scanning electron microscope studies of pollen of cereals and other grasses. *Grana* 12: 79-86.

Andersson, G. (1910) Swedish climate in the Late-Quaternary Period. In: *Die Veranderlingen des Klimas seit dem maximum der Letzten Eiszeit,* 247-294, XI Int. Geol. Congr., Stockholm.

Atherden, M.A. (1976) The impact of late prehistoric cultures on the vegetation of the North York Moors. *Trans. Inst. Br. Geogr.* New Series 1:3: 284-300.

Bailey, J. & G. Culley (1805) *General view of the Agriculture of Northumberland, Cumberland and Westmorland.* London, 3rd Edition (facsimile ed. 1972 by Graham, Newcastle).

Bainbridge, T.H. (1943) Land utilisation in Cumbria in the mid-19thC as revealed by tithe-returns *Trans. Cumberland Westmorland Antiq. and Archaeol. Soc.* NS XLIII, 87-95.

Bannister, P. (1966) Biological Flora of the British Isles No. 102 *Erica tetralix* L. *J. Ecol.* 54: 795-813.

Barber, K.E. (1976) History of Vegetation. In: S.B. Chapman (ed.), *Methods in Plant Ecology,* 5-83. Blackwells Scientific Publications, Oxford.

Barber, K.E. (1977) Recent peat stratigraphy and climatic change. *X INQUA Congress: Abstracts* p.28.

Barrow, G.W.S. (1956) *Feudal Britain.* London, Arnold.

Best, R.H. & J.T. Coppock (1962) *The Changing Use of Land in Britain.* London, Faber.

Beug, H.-J. (1961) *Leitfaden der Pollenbestimmung für Mitteleuropa und angrenzende Gebiete. LI.* Stuttgart, Fischer.

Birks, H.H. (1970) Studies in the vegetational history of Scotland. I: A pollen diagram from Abernethy Forest, Inverness-shire. *J. Ecol.* 58: 827-846.

Birks, H.J.B. (1965) Pollen analytical investigations at Holcroft Moss, Lancashire, and Lindow Moss, Cheshire. *J. Ecol.* 53: 299-314.

Birley, A.R. (1963) *Hadrian's Wall.* London, HMSO.

Blair, P.H. (1956) *An introduction to Anglo-Saxon England.* Cambridge University Press.

Blair, P.H. (1963) *Roman Britain and Early England, 55 BC to AD 871.* Edinburgh: Nelson.

Blytt, A. (1876) *Essay on the immigration of the Norwegian flora during alternating rainy and dry periods.* Christiana: Cammermeye.

Boate, G. (1652) *Ireland's Naturall History.* London.

Boatman, D.J. (1977) Observations on the growth of *Sphagnum cuspidatum* in a bog pool on the Silver Flowe National Nature Reserve. *J. Ecol.* 65: 119-126.

Boatman, D.J. & W. Armstrong (1968) A bog type in north-west Scotland. *J. Ecol.* 56: 129-141.

Boatman, D.J. & R.W. Tomlinson (1973) The Silver Flowe I. Some structural and hydrological features of Brishie Bog and their bearing on pool formation. *J. Ecol.* 61: 633-666.

Braun-Blanquet, J. (1932) *Plant Sociology: the study of plant communities.* McGraw-Hill, New York.

Brightman, F.H. & B.E. Nicholson (1966) *The Oxford Book of Flowerless Plants.* Oxford University Press. 208pp.

Britton, C.E. (1937) A meteorological chronology to AD 1450. *Geophys. Memoir* No. 70, Met. Office, London.

Brookes, D. & K.W. Thomas (1967) The Distribution of Pollen Grains on Microscope Slides Part I: The Non-Randomness of the Distribution. *Pollen et Spores* 9: 621-629.

Brooks, C.E.P. & J. Glasspoole (1928) *British Floods and Droughts.* Benn.

Brown, C.A. (1960) *Palynological Techniques.* Baton Rouge, Louis. 188pp. Private pub.

Casparie, W.A. (1969) Bult-und Schlenkenbildung in Hochmoortorf: (zur Frage des Moorwachstums-Mechanismus). *Vegetatio, Acta Geobotanica* 19: 146-180.

Casparie, W.A. (1972) *Bog development in Southeastern Drenthe (The Netherlands).* The Hague: Junk.

Chapman, S.B. (1964a) The Ecology of Coom Rigg Moss, Northumberland. I Stratigraphy and Present Vegetation. *J. Ecol.* 52: 299-313.

Chapman, S.B. (1964b) The Ecology of Coom Rigg Moss, Northumberland. II The Chemistry of Peat Profiles and the Development of the Bog System. *J. Ecol.* 52: 315-321.

Chapman, S.B. (1965) The Ecology of Coom Rigg Moss, Northumberland. III Some Water Relations of the Bog System. *J. Ecol.* 53: 371-384.

Chapman, S.B., J. Hibble & C.R. Rafarel (1975a) Net aerial production by *Calluna vulgaris* on a lowland heath in Britain. *J. Ecol.* 63: 233-258.

Chapman, S.B., J. Hibble & C.R. Rafarel (1975b) Litter accumulation under *Calluna vulgaris* on a lowland heath in Britain. *J. Ecol.* 63: 359-272.

Clapham, A.R. & H. Godwin (1948). Studies in the Post-Glacial History of British Vegetation. VIII: Swamping surfaces in peats of the Somerset Levels. IX: Prehistoric trackways in the Somerset Levels. *Phil. Trans. R. Soc. B* 233: 233-273.

Clapham, A.R., T.G. Tutin & E.F. Warburg (1962) *Flora of the British Isles.* Cambridge University Press, London. 2nd edition.

Clements, F.E. (1916) Plant succession: an analysis of the development of vegetation. *Carnegie Inst. Washington Publ.* 242: 1-512 Repub. 1928.

Clymo, R.S. (1963) Ion exchange in *Sphagnum* and its relation to Bog Ecology. *Ann. Bot.* 27: 309-324.

Clymo, R.S. (1965) Experiments on breakdown of *Sphagnum* in two bogs. *J. Ecol.* 53: 747-758.

Clymo, R.S. (1970) The growth of *Sphagnum:* methods of measurement. *J. Ecol.* 58: 13-51.

Clymo, R.S. (1973). The growth of *Sphagnum:* some effects of environment. *J. Ecol.* 61: 849-870.

Clymo, R.S. & E.J.F. Reddaway (1974) Growth rate of *Sphagnum rubellum* Wils. on Pennine blanket bog. *J. Ecol.* 62: 191-196.

Coles, J.M., B.J. Orme, F.A. Hibbert & R.A. Jones (1975) *Somerset Levels Papers, 1.* Somerset Levels Project.

Coles, J.M., B.J. Orme, F.A. Hibbert *et al* (1976) *Somerset Levels Papers, 2.* Somerset Levels Project.

Collingwood, R.G. (1933) An introduction to the Prehistory of Cumberland, Westmorland and Lancashire north of the Sands. *Trans. C.W.A.A.S.* NS 23, 163-200.

Conway, V.M. (1948) Von Post's work on climatic rhythms. *New Phytol.* 47: 220-237.

Conway, V.M. (1954) Stratigraphy and Pollen Analysis of Southern Pennine Blanket Peats. *J. Ecol.* 42: 117-147.

Coppock, J.T. (1971) *An Agricultural Geography of Great Britain.* London: Bell.

Cunliffe, B. (1974) The Iron Age. In: C. Renfrew (ed.), *British Prehistory: a new outline:* 233-262.

Dansgaard, W., S.J. Johnsen, N. Reem, N. Gundestrup, H.B. Clausen & C.U. Hammer (1975) Climatic changes, Norsemen and modern man. *Nature* 255: 24-28.

Davies, G. & J. Turner (1979) Pollen diagrams from Northumberland. *New Phytol.* 82: 783-804.

Day, J.B.W. (1970) *Geology of the Country around Bewcastle.* NERC; I.G.S., HMSO. 357pp.

Dickinson, W. (1852) On the farming of Cumberland. *Jl. R. Agric. Soc.* 13: 207-300.

Dickinson, W. (1975) Recurrence surfaces in Rusland Moss, Cumbria (formerly North Lancashire). *J. Ecol.* 63: 913-935.

Dickson, C.A. (1970) The study of plant macrofossils in British Quaternary deposits. In: D. Walker & R.G. West (eds.), *Studies in the Vegetational History of the British Isles:* 233-254.

Dickson, J.H. (1973) *Bryophytes of the Pleistocene.* Cambridge University Press.

Dimbleby, G.W. (1967) *Plants and Archaeology.* London: Baker.

Donaldson, A.M. & J. Turner (1977) A pollen diagram from Hallowell Moss, near Durham City, UK. *J. Biogeog.* 4: 25-33.

Duce, R.A., J.T. Wasson, J.W. Winchester & F. Burns (1963) Atmospheric Iodine, Bromine and Chlorine. *J. Geophys. Research* 68: 3943-3947.

Duce, R.A., J.W. Winchester & T.W. van Nahl (1965) Iodine, Bromine and Chlorine in the Hawaiian Marine Atmosphere. *J. Geophys. Research* 70: 1775-1799.

Duncan, U.K. (1962) Illustrated Key to *Sphagnum* mosses. *Trans. & Proc. Bot. Soc. Edinb.* 39: 290-301.

Du Rietz, G.E. (1921) *Zur methodologischen Grundlage der modernen Pflanzensoziologie.* Holzhausen, Wien.

Du Rietz, G.E. (1930) Classification and nomenclature of vegetation. *Svensk. Bot. Tidskr.* 24: 489-503.

Du Rietz, G.E. (1950) Phytogeographical excursion to the Ryggmossen Mire near Uppsala. *7th Internat. Bot. Congress, Excursion Guide AIIb3.* Stockholm.

Dutch, R.A. (1962) *Roget's Thesaurus of English Words and Phrases.* Longmans.

Duvigneaud, P. (1949) Classification phytosociologique des toubieres de l'Europe. *Bull. Soc. Roy. Bot. Belgique.* 81: 59-122.

Elliot, G. (1956) *Some aspects of the changing agricultural landscape in Cumberland.* MA thesis: Liverpool University.

Elliott, G. (1959) The system of cultivation and evidence of enclosure in the Cumberland open fields in the 16th century. *Trans. CWAAS* NS 59: 85-104.

Elliot, G. (1973) Field Systems of Northwest England. In: A.H.R. Baker & R.A. Butlin (eds.), *Studies of field systems in the British Isles.* Cambridge University Press, 42-92.

Erdtman, G. (1929) Some aspects of the post-glacial history of British forests. *J. Ecol.* 17: 112-126.

Ernst, O. (1934) Zur Geschickte der Moore, Marschen und Wälder Nordwest-Deutschland.

IV: Untersuchungen in Nordfriesland. *Schr. Natur. Ver. Schlesw. Holst.* 20: 209-334.
Erdtman, G. (1943) *An Introduction to Pollen Analysis.* Waltham, Mass: Chronica Botanica.
Erdtman, G., B. Berglund & J. Praglowski (1961) An introduction to a Scandinavian Pollen Flora. *Grana Palynologica* 2: 3-92.
Erdtman, G., J. Praglowski & S. Nilsson (1963) *An introduction to a Scandinavian Pollen Flora, Vol. II.* Stockholm: Almquist & Wiksell.
Eurola, S. (1962) Uber die Regionale Einteilung der Südfinnischen Moore. *Ann. Bot. Soc. 'Vanamo'* 33:2: 1-243.
Faegri, K. & J. Iversen (1964) *Textbook of Pollen Analysis.* Copenhagen: Munksgaard.
Faegri, K. & J. Iversen (1975) *Textbook of Pollen Analysis.* Oxford: Blackwell Scientific Publications, 3rd edition.
Feachem, R.W. (1973) Ancient agriculture in the Highlands of Britain. *Proc. Prehist. Soc.* 39: 332-353.
Fearnsides, M. (1938) Graphic keys for the identification of Sphagna. *New Phytol.* 37: 409-424.
Ferguson, R.S. (1890) *A History of Cumberland.* London: Elliot Stock.
Forrest, G.I. (1971) Structure and production of North Pennine blanket bog vegetation. *J. Ecol.* 59: 453-479.
Forrest, G.I. & R.A.H. Smith (1975) The productivity of a range of blanket bog vegetation types in the northern Pennines. *J. Ecol.* 63: 173-202.
Fox, C. (1959) *The Personality of Britain.* Cardiff: Nat. Mus. of Wales 4th Edition (1st Edition: 1932).
Fraser, G.M. (1971) *The Steel Bonnets. The story of the Anglo-Scottish Border Reivers.* Barrie & Jenkins.
Franklin, T.B. (1953) *British grasslands from the earliest times to the present day.* Faber, London.
Frenzel, B. (1966) Climatic change at the Atlantic-Sub-Boreal transition in the northern Hemisphere: botanical evidence. *Proc. Int. Symp. R. Met. Soc. Lond.: World Climate from 800-0 BC:* 99-122.
Früh, J.J. & C. Schröter (1904) *Die Moore der Schweiz mit Berücksichtigung der gesamten Moorfrage.* Zurich.
Fussell, G.E. (1964) The Grasses and grassland cultivation of Britain. *J. Br. Grassld. Soc.* 19: 212-217.
Ganong, W.F. (1897) Upon raised peat bogs in the province of New Brunswick. *Proc. & Trans. Roy. Soc. of Canada,* 2nd series, 3 Pub. Montreal.
Garlick, T. (1972) *Romans in the Lake Counties.* Clapham; Dalesman Books.
Geel, B. van (1976) *A palaeoecological study of Holocene peat bog sections, based on the analysis of pollen, spores and macro and microscopic remains of fungi, algae, cormophytes and animals.* Thesis, University of Amsterdam (also published in *Vegetatio,* 1977).
George, D. (1953) *England in Transition.* London: Penguin.
Gimingham, C.H. (1960) Biological flora of the British Isles: *Calluna vulgaris* (L.) Hull. *J. Ecol.* 48: 458-483.
Gimingham, C.H. (1972) *Ecology of the Heathlands.* London: Chapman & Hall.
Gimingham, C.H., G.R. Miller, L.M. Sleigh & L.M. Milne (1961) The ecology of a small bog in Kinlochewe Forest, Wester Ross. *Trans. Bot. Soc. Edinb.* 39: 125-147.
Glob, P.V. (1969) *The Bog People.* Faber & Faber.
Godwin, H. (1934) Pollen analysis: an outline of the problems and potentialities of the method: Part I: Technique and interpretation. *New Phytol.* 33: 278-305.
Godwin, H. (1940) Pollen analysis and forest history of England and Wales. *New Phytol.* 39: 370-400.

Godwin, H. (1940) Studies of the post-glacial history of British vegetation III: Fenland pollen diagrams. *Phil. Trans. Roy. Soc. B.* Vol. 230, 239.
Godwin, H. (1941) Studies of the Post-glacial history of British vegetation IV: Correlations in the Somerset Levels. *New Phytol.* 40: 108.
Godwin, H. (1946) The relationship of bog stratigraphy to climatic change and archaeology. *Proc. prehist. Soc.* 12: 1-11.
Godwin, H. (1954) Recurrence Surfaces. *Danm. Geol. Unders.* II R 80: 22-30.
Godwin, H. (1956) *History of the British Flora.* Cambridge University Press.
Godwin, H. (1960) Radiocarbon dating and Quaternary history in Britain. *Proc. R. Soc.* B 153: 287-320.
Godwin, H. (1967) The ancient cultivation of hemp. *Antiquity* XLI: 42-50.
Godwin, H. (1975) *History of the British Flora:* 2nd edition. Cambridge University Press.
Godwin, H. & V.M. Conway (1939) The Ecology of a raised bog near Tregaron, Cardiganshire. *J. Ecol.* 27: 315-359.
Godwin, H., D. Walker & E.H. Willis (1957) Radiocarbon dating and post-glacial vegetational history: Scaleby Moss. *Proc. R. Soc.* B 147: 352-366.
Goode, D. (1970) *Ecological studies on the Silver Flowe Nature Reserve.* Ph.D. thesis, University of Hull.
Goodman, G.T. & D.F. Perkins (1968) The role of mineral nutrients in *Eriophorum* communities. IV Potassium supply as a limiting factor in a *E.vaginatum* community. *J. Ecol.* 56: 685-696.
Gore, A.J.P. & C. Urquhart (1966) The effects of waterlogging on the growth of *Molinia caerulea* and *Eriophorum vaginatum. J. Ecol.* 54: 617-633.
Gorham, E. (1949) Some chemical aspects of a peat profile. *J. Ecol.* 37: 24-27.
Gorham, E. (1953) A note on the acidity and base status of raised and blanket bogs. *J. Ecol.* 41: 153-156.
Gorham, E. (1953) Some early ideas concerning the nature, origin and development of peat lands. *J. Ecol.* 41: 257-274.
Gorham, E. (1957) The Development of Peat Lands. *Quart. Rev. Biol.* 32: 145-166.
Gorham, E. & W.H. Pearsall (1956) Acidity, Specific Conductivity and Calcium content of some bog and fen waters in northern Britain. *J. Ecol.* 44: 129-141.
Graaf, Fr. de (1956) Studies on Rotatoria and Rhizopoda from the Netherlands. I. Rotatoria and Rhizopoda from the 'Grote Huisven'. *Biol. Jaarb.* 23: 145-217.
Granlund, E. (1932) De Svenska Högmossarnas geologi. *Sver. geol. Unders.* C 26: 1-193.
Gray, J. (1965) Extraction Techniques. In: B. Kummel & D. Raup (eds.), *Handbook of Paleontological Techniques,* 530-587. San Francisco: Freeman.
Green, B.H. (1965) *Some studies of Water/Peat/Plant relationships with Special Reference to Wybunbury Moss, Cheshire.* Unpub. Ph.D. thesis, University of Nottingham.
Green, B.H. (1968) Factors influencing the spatial and temporal distribution of *Sphagnum imbricatum* Hornsch ex Russ. in the British Isles. *J. Ecol.* 56: 47-58.
Green, F.H.W. (1964) A map of annual average potential water deficit in the British Isles. *J. Appl. Ecol.* 1: 151-158.
Grigg, D.B. (1967) The Changing Agricultural Geography of England: a commentary on the sources available for the reconstruction of the Agricultural Geography of England, 1770-1850. *Trans. Inst. Br. Geogr.* 41: 73-96.
Grospeitch, T. (1953) Rhizopodenanalytische Untersuchungen an Mooren Ostholsteins. *Arch. Hydrobiol.* 47: 321-452.
Grosse-Brauckmann, G. (1963) Zur Artenzusammensetzung von Torfen. *Ber. Deutsch. bot. Gesell.* 76: 22-35.
Hansen, B. (1966) The Raised Bog Draved Kongsmose. *Bot. Tidsskr.* 62: 2-3: 146-185.
Harley, J.B. (1962-1963) A Guide to Ordnance Survey Maps as Historical Sources. I The

One Inch to One Mile Maps of England and Wales III Six Inch and Twenty Five Inch IV Town Plans and Small Scale Maps. *Am. Hist.* 5, 6 and 7.

Harley, J.B. (1972) *Maps for the local historian. A guide to the British sources.* London: National Council of Social Service (for the Standing Conference for Local History).

Harley, J K. & E.W. Yemm (1942) Ecological aspects of peat accumulation: I Thornton Mire, Yorkshire. *J. Ecol.* 30: 17-56.

Heal, O.W. (1961) The distribution of testate amoebae (Rhizopoda: Testacea) in some fens and bogs in northern England. *J. Linn. Soc. (Zoo.)* 44: 369-382.

Heal, O.W. (1962) The abundance and microdistribution of testate amoebae (Rhozopoda: Testacea) in *Sphagnum. Oikos* 13: 35-47.

Heal, O.W. (1964) Observations on the seasonal and spatial distribution of Testacea (Protozoa: Rhizopoda) in *Sphagnum. J. Anim. Ecol.* 33: 395-412.

Hibbert, F.A., V.R. Switsur & R.G. West (1971) Radiocarbon dating of Flandrian pollen zones at Red Moss, Lancashire. *Proc. R. Soc.* B: 177: 161-176.

Hibbert, F.A. & V.R. Switsur (1976) Radiocarbon dating of Flandrian pollen zones in Wales and Northern England. *New Phytol.* 77: 793-807.

Highham, N.J. & G.B.D. Jones (1976) Frontier, Forts and Farmers (Cumbrian Aerial Survey 1974-5). *Arch. Journal* 132: 16-53.

Hogg, R. (1972) Factors which have affected the spread of early settlement in the Lake Counties. *Cumb. & West. Ant. & Arch. Soc. (Trans.)* LXXII: 1-36.

Hodgson, W. (1898) *Flora of Cumberland.* Meals: Carlisle.

Houseman, J. (1800) *A topographical description of Cumberland Westmoreland, Lancashire and part of the West Riding of Yorkshire.* Carlisle.

Hutchinson, W. (1794) *The history of the county of Cumberland, and some place adjacent . . . comprehending the local history of the county, its antiquities, the origin, geneology and present state of the principal families etc.* Carlisle: Jollie, 2 vol. (reprinted 1974 by E.P. Publishing, Wakefield).

Hunter-Blair, P. (1963) *Roman Britain and Early England, 55BC-871AD.* London: Nelson.

Ingram, H.A.P. (1967) Problems of hydrology and plant distribution in mires. *J. Ecol.* 55: 711-725.

Isoviita, P. (1966) Studies on *Sphagnum* L. 1. Nomenclatural revision of the European taxa. *Ann. Bot. Fenn.* 3: 199-264.

Jefferson, S. (1838) *The History and Antiquities of Carlisle: with an account of the castle, Gentlemen's Seats and Antiquities in the vicinity etc.* Carlisle: by S. Jefferson & Whittaker & Co, London.

Johnsen, S.J., W. Dansgaard, H.B. Clausen & C.C. Langway (1972) Oxygen Isotope profiles through the Antarctic and Greenland Ice Sheets. *Nature* 235: 429-434.

Jollie, F. (1811) *Jollie's Cumberland Guide and Directory; containing a descriptive tour through the county, etc.* Carlisle.

Jones, G.P. (1956) The poverty of Cumberland and Westmorland. *Trans CWAAS* LV: 198-208.

Kershaw, K.A. (1964) *Quantitative and dynamic ecology.* London, Arnold.

Kershaw, K.A. (1973) *Quantitative and dynamic plant ecology*, 2nd edition. London, Arnol

King, W. (1685) On the bogs and loughs of Ireland. *Phil. Trans.* 15: 948.

Katz, N.J. (1926) *Sphagnum* bogs of central Russia: phytosociology, ecology and succession. *J. Ecol.* 14: 177-202.

Koestler, A. (1969) *The Act of Creation.* Hutchinson.

Kulczynski, S. (1949) Peat Bogs of Polesie. *Mem. de l'Acad. Polon. des Sciences et des Lettres,* Série B: 1-356.

Ladurie, E. Le Roy. (1972) *Times of Feast, Times of Famine. A history of climate since the year 1000.* London: Allen & Unwin.

Lamb, H.H. (1965a) The Early Medieval Warm Epoch and its sequel. *Palaeogeography, Palaeoclimatology, Palaeoecology* 1: 13-37.
Lamb, H.H. (1965b) Britain's Changing Climate. In: C.G. Johnson & L.P. Smith (eds.), *The Biological Significance of Climatic Changes in Britain.* Institute of Biology Symposium No. 14. Academic Press: 3-31.
Lamb, H.H. (1966) *The Changing Climate.* London, Methuen.
Lamb, H.H. (1977a) The late Quaternary history of the climate of the British Isles. In: F.W. Shotton (ed.), *British Quaternary Studies: Recent Advances.*
Lamb, H.H. (1977b) *Climate: Present, Past and Future. Vol. 2: Climatic history and the future.* London, Methuen.
Lein, A. & N. Schwartz (1951) Ceric sulfate-arsenious acid reaction in microdetermination of iodine. *Analyt. Chem.* 23(10): 1507-1510.
Lewis, F.J. (1904) Geographical distribution of the vegetation of the basins of the rivers Eden, Tees, Wear and Tyne. Parts I & II. *Geogr. J.* 23.
Lock, M.A., P.M. Wallis & H.B.N. Hynes (1977) Colloidal organic carbon in running waters. *Oikos* 29: 1-4.
Lundqvist, B. (1962) Geological Radiocarbon Datings from the Stockholm station. *Sver. Geol. Unders.* C 589: 3-23.
Manley, G. (1959) Temperature trends in England 1698-1957. *Archiv für Met. Geophys. und Biokl.* Serie B: 413-433.
Manley, G. (1965) Possible climate agencies in the development of post-glacial habitats. *Proc. R. Soc.* B 161: 363-375.
Manley, G. (1966) The problem of the climatic optimum: the contribution of glaciology. *Proc. Internat. Symp. R. Met. Soc.: World Climate 8000-0 BC:* 34-39.
Manley, G. (1974) Central England temperatures: monthly means 1659-1973. *Quart. J. Roy. Met Soc.* 100: 389-405.
Manning, W.H. (1975) Economic influences on land use in the military areas of the Highland Zone during the Roman period. In: C.B.A. Research Report No. 11: *The effect of man on the landscape: The Highland Zone:* 112-116.
Mattson, E. & E. Koulter-Andersson (1954) Geochemistry of a raised bog. *Köngl. Lantbrukshögskolars Ann.* 21: 321-366.
McKerrell, H. (1975) Correction Procedures for C-14 Dates and Appendix I: Conversion Tables. In: T. Watkins (ed.), *Radiocarbon: Calibration and Prehistory.* Edinburgh University Press: 47-100 and 110-127.
McVean, D.N. & D.A. Ratcliffe (1962) *Plant Communities of the Scottish Highlands.* Monographs of the Nature Conservancy No. 1 HMSO London.
Millington, R.J. (1954) *Sphagnum* bogs of the New England Plateau, New South Wales. *J. Ecol.* 42: 328-344.
Millward, R. & A. Robinson (1972) *Cumbria.* London: Macmillan.
Mitchell, A. (1974) *A field guide to the trees of Britain and Northern Europe.* London: Collins.
Mitchell, G.F. (1965) Littleton Bog, Tipperary: an Irish agricultural record. *J. Roy. Soc. Antiquaries of Ireland* 95: 121-132.
Miyake, Y. & S. Tsunogai (1963) Evaporation of Iodine from the Ocean. *Jl. Geochem. Res.* 68: 3989-3993.
Moore, P.D. & D.J. Bellamy (1974) *Peatlands.* Elek: London.
Morrison, M.E.S. (1955) The water balance of the raised bog. *Irish Nat. J.* XI No. 11: 303-308.
Morrison, M.E.S. (1959) The Ecology of a raised bog in Co. Tyrone, Northern Ireland. *Proc. R. Ir. Acad.* 60B: 291-308.
Moss, C.E. (1911) The Upland Moors of the Pennine Chain. In: A.G. Tansley (ed.), *Types of British Vegetation,* 266-282.

Myers, A.R. (1963) *England in the Late Middle Ages.* London: Penguin Books, 2nd edition.
Nicholas, F.J. & H. Glasspoole (1931) General monthly rainfall over England and Wales 1727-1931. *Brit. Rainf.* 299-306.
Nicolson, J. & R. Burn (1777) *The History and Antiquities of Cumberland and Westmorland.* London: Strahen & Cadell.
Nillson, T. (1964) Standardpollendiagramme und C^{14} Datierungen aus dem Agerods mosse in Mittleren Schonen. *Lunds Univ. Arsskr.* N.F. (2) 59:7: 1-52.
Nillson, T. (1964) Entwicklungsgeschichliche Studien im Ageröds Mose, Schonen. *Lunds Univ. Arsskr.* 59: 8.
Olaussen, E. (1957) Das Moor Roshultsmyren. *Lunds Univ. Arsskr.* NF Adv. 2. 53: 1-72.
Oldfield, F. (1960) Studies in the post-glacial history of British Vegetation: Lowland Lonsdale. *New Phytol.* 59: 192-217.
Oldfield, F. (1963) Pollen analysis and Man's role in the Ecological History of the South-east Lake District. *Geogr. Ann.* 45: 23-40.
Oldfield, F. (1969) Pollen analysis and the history of land-use. *Advmt. Sci. Lond.* 25: 298-311.
Oldfield, F., R. Thompson & K.E. Barber (1978) Changing Atmospheric Fallout of Magnetic Particles Recorded in Recent Ombrotrophic Peat Sections. *Science* 199: 679-80.
Oldfield, F., P.G. Appleby, R.S. Cambray, J.D. Eakins, K.E. Barber, R.W. Battarbee, G.W. Pearson & J.M. Williams (1979) ^{210}Pb, ^{137}Cs and ^{239}Pu profiles in ombrotrophic peat. *Oikos* 33: 40-45.
Ordnance Survey (1860-63) *Ordnance Survey Record Book – Kirklinton Parish* 1-27.
Osvald, H. (1923) Die Vegetation des Hochmoores Komosse. *Svensk. Växtsoc. Sallsk. Handl.* 1.
Osvald, H. (1925a) Zur Vegetation der Ozeanischen Hochmoore in Norwegen. *Svenska Vaxsociologiska Sallskapets Handlingar* VII: 106p; 16pl.
Osvald, H. (1925b) Die Hochmoortypen Europas. *Veröffentlichungen des Geobotanischen Institutes Rubel in Zurich,* 3. Heft: Festschrift Carl Schroter: 707-723.
Osvald, H. (1933) Vegetation of the Pacific Coast bogs of North America. *Acta Phytogeogr. Suecica* 5: 1-33.
Osvald, H. (1937) *Myrar och Myrodling.* Stockholm: Kooperativa Förbundets Bokförlag.
Osvald, H. (1949) Notes on the vegetation of British and Irish Mosses. *Acta Phytogeog. Suecica.* 26: 7-62.
Osvald, H. (1950) The raised bog Komosse. *7th Internat. Bot. Congress, Excursion Guide AIIb2.* Stockholm.
Overbeck, F. (1946) Studien zur Hochmoorentwicklung in Niedersachsen und die Bestimmung der Humifizierung der Stratigraphisch-pollen analytischen Mooruntersuchungen. *Planta* 35: 1-56.
Overbeck, F. (1963) Aufgaben botanisch-geologischer Moorforschung in Nordwestdeutschland. *Ber. Deutsch. Bot. Gesell.* 76: 2-12.
Overbeck, F. & I. Griez (1954) Mooruntersuchungen zur Rekurrenzflächenfrage und Siedlungsgeschichte in der Rhön. *Flora.* 141: 51-94.
Overbeck, F. & H. Happach (1957) Über das Wachstum und den Wasserhaushalt einiger Hochmoorsphagnen. *Flora* 144: 336-402.
Parry, M.L. (1975) Secular climatic change and marginal agriculture. *Trans. Inst. Br. Geogr.* 64: 1-14.
Paulson, B. (1952) Some rhizopod associations in a Swedish mire. *Oikos* 4: 151-165.
Pearsall, W.H. (1938) The soil complex in relation to plant communities. III Moorland bogs. *J. Ecol.* 26: 298-315.
Pearsall, W.H. (1941) The 'mosses' of the Stainmore district. *J. Ecol.* 29: 161-175.
Pearsall, W.H. (1956) Two blanket-bogs in Sutherland. *J. Ecol.* 44: 493-516.

Pearsall, W.H. (1963) The development of ecology in Britain. *J. Ecol.* (Jubilee Suppl.) 51: 1-12.
Pearsall, W.H. & E.M. Lind (1941) A note on a Connemara bog type. *J. Ecol.* 29: 62-68.
Pennington, W. (1970) Vegetation history in the north-west of England: a regional synthesis. In: D. Walker & R.G. West (eds.), *Studies in the vegetational history of the British Isles:* 41-80.
Pennington, W. & J.P. Lishman (1971) Iodine in lake sediments in Northern England and Scotland. *Biol. Rev.* 46: 279-313.
Perring, F.H. & S.M. Walters (eds.) (1962) *Atlas of the British Flora.* London: Nelson.
Pigott, C.D. & M.E. Pigott (1959) Stratigraphy and Pollen Analysis of Malham Tarn and Tarn Moss. *Field Studies* 1: 1-18.
Pigott, C.D. & M.E. Pigott (1963) Late-glacial and Post-glacial deposits at Malham, Yorkshire. *New Phytol.* 62: 317-324.
Pinnock, W. (1822) *The History and Topography of Cumberland.* London.
Post, L. von & R. Sernander (1910) Pflanzen-physiognomische Studien auf Torfmooren in Närke. *XI International Geological Congress: Excursion Guide No. 14 (A7)* Stockholm. 48p.
Proctor, M.C.F. (1955) Key to the British species of *Sphagnum. Trans. Brit. Bryol. Soc.* 2: 552-560.
Rankin, W.M. (1911) The Lowland Moors of Lonsdale. In A.G. Tansley (ed.), *Types of British Vegetation:* 247-59.
Ratcliffe, D.A. (1964) Mires and Bogs. In: J.H. Burnett (ed.), *The Vegetation of Scotland:* Edinburgh, Oliver & Boyd: 426-478.
Ratcliffe, D.A. & D. Walker (1958) The Silver Flowe, Galloway, Scotland. *J. Ecol.* 46: 407-445.
Rennie, R. (1807-10) *Essays on the Natural History and Origin of Peat Moss, parts I-IX.* Edinburgh.
Richards, P.W. & E.C. Wallace (1950) An annotated list of British Mosses. *Trans. Br. Bryol. Soc.* 1 Suppl.
Richmond, I.A. (1958) *Roman and Native in North Britain.* London: Nelson.
Roberts, B.K., J. Turner & P.F. Ward (1973) Recent forest history and land use in Weardale, Northern England. In: H.J.B. Birks & R.G. West (eds.), *Quaternary Plant Ecology:* 207-221. Oxford: Blackwell Scientific Publications.
Romanov, V.V. (1966) *Hydrophysics of Bogs.* Leningrad – translated by Israel Program for Scientific Translations, Jerusalem 1968.
Rovner, I. (1971) Potential of Opal Phytoliths for use in palaeoecological reconstruction. *Quaternary Research* 1:3: 343-359.
Rybnickova, E. & K. Rybnicek (1971) The determination and elimination of local elements in pollen specta from different sediments. *Rev. Palaeobot. & Palynol.* 11: 165-176.
Rycroft, D.W., D.J.A. Williams & H.A.P. Ingram (1975a) The transmission of water through peat. I Review. *J. Ecol.* 63: 535-556.
Rycroft, D.W., D.J.A. Williams & H.A.P. Ingram (1975b) The transmission of water through peat II Field experiments. *J. Ecol.* 63: 557-568.
Ryle, G. (1969) *Forest Service.* Newton Abbot: David & Charles.
Sagar, G.R. & J.L. Harper (1964) Biological flora of the British Isles: No. 95. *Plantago major, P. media* and *P. lanceolata. J. Ecol.* 52: 189-221.
Schneekloth, H. (1963) Die Rekurrenzflache – eine Zeitgleiche Bildung innerhalb eines Hochmoores? *Ber. Deutsch. Bot. Gesell.* 76: 14-16.
Schneekloth, H. (1965) Die Rekurrenzflache im Grossen Moor Bei Gifhorn – eine Zeitgleiche Bildung? *Geol. Jb.* 83: 477-496.

Scott, J.G. (1966) *South-west Scotland.* London: Heinemann (Regional Archaeologies).
Seaward, M.R.D. (1976) *The Vindolanda Environment.* Haltwhistle, Barcombe Publications.
Sernander, R. (1908) On the evidence of Postglacial changes of climate furnished by the peat-mosses of northern Europe. *Geol. Fören. Förh.* 30: 465-478.
Sernander, R. (1909) De scanodaniska torfmossarnas stratigrafi. *Geol. Fören. Förh.* 31: 423-448.
Sernander, R. (1910a) Ausstellung zur Belauchtung der Entwicklung – geschichte der schwedischen Torfmoore. *Compt. Rend. XI Cong. Geol. Internat. 203.*
Sernander, R. (1910b) Excursion B3: Örsmossen. *Compt. Rend. XI Cong. Geol. Internat. 1292.*
Sernander, R. & K. Kjellmark (1895) Eine Torfmooruntersuchung aus dem nördlichen Nerike. *Bull. Geol. Inst. Univ. Upsala,* 2.
Shacklette, H.T. & M.E. Cuthbert (1967) Iodine content of plant groups as influenced by variation in rock and soil type. *Geol. Soc. Am. Special Paper* 90: 31-46.
Shimwell, D.W. (1971) *Description and Classification of Vegetation.* Sidgwick & Jackson, 322pp.
Sjörs, H. (1946) The mire vegetation of the Upper Långan District in Jamtland. *Ark. Bot. Uppsala* 33: 1-96.
Sjörs, H. (1948) Myrvegetation i Bergslagen. *Acta Phytogeogr. Suecica,* 21: 16-299.
Sjörs, H. (1950) On the relation between vegetation and electrolytes in North Swedish mire waters. *Oikos* 2:2: 241-258.
Sjörs, H. (1950) Regional studies in Swedish mire vegetation. *Bot. Notiser* 2: 173-222.
Slicher van Bath, B.H. (1963) *The Agrarian History of Western Europe A.D. 500-1850.* London: Arnold.
Smith, A.G. (1959) The mires of south-western Westmorland: stratigraphy and pollen-analysis. *New Phytol.* 58: 105-127.
Sparks, B.W. & R.G. West (1972) *The Ice Age in Britain.* London: Methuen.
Spring, D. (1955) A great agricultural estate: Netherby under Sir James Graham, 1820-1845. *Agric. History* XXIX:2: 73-81.
Sykes, J.B. (1976) *The Concise Oxford Dictionary of Current English.* Oxford University Press.
Tallis, J.H. (1962) The identification of *Sphagnum* spores. *Trans. Br. Bryol. Soc.* 4: 209-213.
Tallis, J.H. (1964) Studies on Southern Pennine Peats. III: The behaviour of *Sphagnum. J. Ecol.* 52: 345-353.
Tallis, J.H. (1973) The terrestrialization of lake basins in North Cheshire, with special reference to the development of a 'Schwingmoor' structure. *J. Ecol.* 61: 537-567.
Tansley, A.G. (1939) *The British Islands and their Vegetation.* Cambridge: University Press.
Tansley, A.G. (1949) *Britain's Green Mantle.* London: Allen & Unwin.
Thomson, D. (1950) *England in the Nineteenth Century.* London: Penguin.
Tinsley, H. (1976) Cultural influences on Pennine vegetation with particular reference to North Yorkshire. *Trans. Inst. Br. Geogr. New Series,* 1:3: 310-322.
Tinsley, H.M. & R.T. Smith (1974) Surface pollen studies across and woodland/heath transition and their application to the interpretation of pollen diagrams. *New Phytol.* 73: 547-565.
Tolonen, K. (1966a) Stratigraphic and rhizopod analyses on an old raised bog, Varrasuo in Hollola, South Finland. *Ann. Bot. Fenn.* 3: 147-166.
Tolonen, K. (1967) Über die Entwicklung der Moore im finnischen Nordkarelien. *Ann. bot. Fenn.* 4.

Tolonen, K. (1968) Zur Entwicklung der Binnenfinnland-Hochmoore. *Ann. bot. Fenn.* 5: 17-33.

Tolonen, K. (1971) On the regeneration of North European Bogs. I Klaukkalan Isosuo in S. Finland. *Acta Agralia Fennica* 123: 143-166.

Tolonen, K. & R. Ruuhijarvi (1976) Standard pollen diagrams from the Salpausselka region of Southern Finland. *Ann. Bot. Fenn.* 13: 155-196.

Troels-Smith, J. (1955) Characterization of unconsolidated sediments. *Danm. geol. Unders. IVR* 3: 10.

Trotter, F.M. & S.E. Hollingworth (1932) *The geology of the Brampton district.* Mem. geol. Surv. Gt. Br.

Turner, J. (1965) A contribution to the history of forest clearance. *Proc. R. Soc. B* 161: 343-392.

Turner, J. (1970) Post-Neolithic disturbance of British vegetation. In: D. Walker & R.G. West (eds.), *Studies in the Vegetational History of the British Isles,* pp.97-116. Cambridge University Press.

Turner, J. (1975) The evidence for land use by prehistoric farming communities: the use of three-dimensional pollen diagrams. In: J.G. Evans, S. Limbrey & H. Cleere (eds.), *The Effect of Man on the Landscape: the Highland Zone.* Council for British Archaeology Research Report No. 11.

Walker, D. (1961) Peat Stratigraphy and Bog Regeneration. *Proc. Linn. Soc.* 172: 29-33.

Walker, D. (1966) The Late Quaternary history of the Cumberland Lowland. *Phil. Trans. R. Soc.* B251: 1-210.

Walker, D. (1970) Direction and rate in some British Post-glacial hydroseres. In: D. Walker & R.G. West (eds.), *Studies in the Vegetational History of the British Isles.* 117-139.

Walker, D. & P.M. Walker (1961) Stratigraphic evidence of regeneration in some Irish bogs. *J. Ecol.* 49: 169-185.

Walker, J. (1772) Account of the Irruption of Solway Moss in December 16, 1771. *Phil. Trans.* LXII: 123-127.

Watt, A.S. (1947) Pattern and process in the plant community. *J. Ecol.* 35: 1-22.

Watson, E.V. (1968) *British Mosses and Liverworts.* Cambridge University Press. 2nd edition.

Weber, C.A. (1900) Über die Moore, mit besondere Berucksichtigung der zwischen Unterweser and Unterelbe liegenden. *Jahresbericht der Manner von Morgenstern* 3: 3-23.

Weber, C.A. (1902) *Über die Vegetation und Entstehung des Hochmoores von Augstumal im Memeldelta mit vergleichenden Ausblicken auf andere Hochmoore der Erde.* Berlin Paul Parey. 252pp.

Weber, C.A. (1908) Aufbau und Vegetation der Moore Norddeutschlands. *Bot. Jb.* (Suppl.) 90: 19-34.

Wein, R.W. (1973) Biological flora of the British Isles *Eriophorum vaginatum* L. *J. Ecol.* 61: 601-615.

West, R.G. (1968) *Pleistocene Geology and Biology.* Longmans: London.

West, R.G. (1970) Pollen Zones in the Pleistocene of Great Britain and their correlation. *New Phytol.* 69: 1179-1183.

West, R.G. (1977) *Pleistocene Geology and Biology:* 2nd edition. London: Methuen.

Wilson, D.R. (1967) *Roman Frontiers of Britain.* London: Heinemann.

Wonnacott, T.H. & R.J. Wonnacott (1972) *Introductory Statistics for Business and Economics.* New York: Wiley.

Woodhead, T.W. (1929) History of the vegetation of the Southern Pennines. *J.Ecol.* 17: 1-34.

Wright, H.E., jr. & H.L. Patten (1963) The pollen sum. *Pollen et Spores,* 5: 445-450.

Ziegler, P. (1966) *The Black Death.* Collins: London.